DETECTION AND DIAGNOSIS OF PLANT DISEASES

Dr. A K Meena
M. Sc. (Ag) Plant Pathology, Ph.D.
Assistant Professor
Department of Plant Pathology,
College of Agriculture Bikaner (Rajasthan)

Dr. S. L. Godara
M. Sc. (Ag) Plant Pathology, Ph. D.
Professor in the Department of Plant Pathology
Director Research, Swami Keshwanand Rajasthan
Agricultural University, Bikaner.

Dr. P N Meena
M. Sc. (Ag) Plant Pathology, Ph. D.
Scientist in Crop Protection Division,
ICAR-CRIJAF, Barrackpore,
Kolkata, West Bengal

SCIENTIFIC PUBLISHERS

Published by

SCIENTIFIC PUBLISHERS (INDIA)
5 A, New Pali Road, P.O. Box 91
Jodhpur 342 001 (INDIA)
E-mail: info@scientificpub.com
Website : www.scientificpub.com

ISBN: 978-93-89184-40-2
eISBN: 978-93-89184-42-6

Printed in India

A Practical Book

DETECTION AND DIAGNOSIS OF PLANT DISEASES

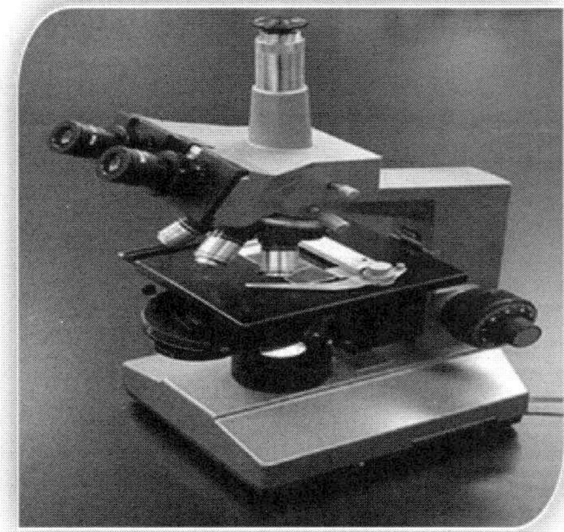

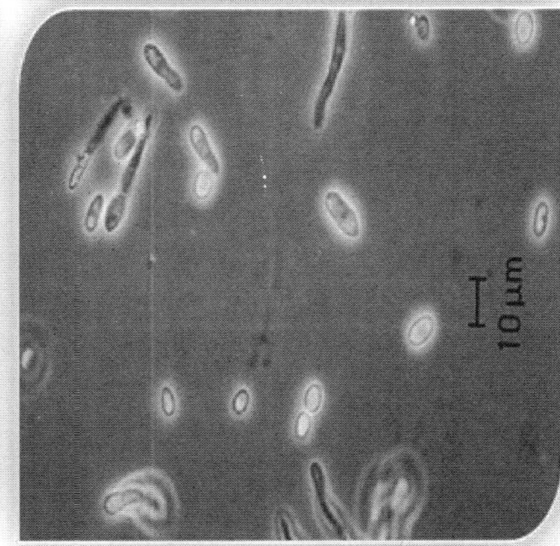

A K Meena
S L Godara
P N Meena

Dr. A K Meena

Dr. A K Meena did his Ph.D. from SKRAU, Bikaner (Rajasthan). He is currently Assistant Professor in the Department of Plant Pathology, College of Agriculture, Bikaner. During the last five years, he has been engaged in research and teaching and published many research papers, articles and book chapters in national and international reputed journals. He is also life member in various Plant Pathological Societies. He has expertise in Plant Pathology, Plant Bacteriology and Plant Virology.

Dr. S. L. Godara

Dr. S. L. Godara, Professor in the Department of Plant Pathology and presently working as Director Research, Swami Keshwanand Rajasthan Agricultural University, Bikaner. He did his M. Sc. (Ag) Plant Pathology in 1986 from Sukhadia University, Udaipur and Ph. D. in 1995 from Rajasthan Agricultural University, Bikaner. Dr. Godara is a Fellow of Indian Society of Mycology and Plant Pathology and Indian Society of Plant Pathologists. He has three decade long standing experience in the field of education, research and technology transfer. He has been working on management of plant diseases of important field crops *viz;* mustard, groundnut, cumin, mothbean, clusterbean and post harvest diseases of citrus, ber and guava fruits and Principal Investigator of five research projects on important crops of the region. He has published more than 50 research papers in national as well as international journals, one book on "Disease management of spices crops" and seven practical manuals.

Dr. P N Meena

Mr. P N Meena is currently scientist in Crop Protection Division, ICAR-CRIJAF, Barrackpore, Kolkata, West Bengal. He did M. Sc. from MPUAT, Udaipur (Rajasthan) with Junior Research Fellow (JRF-ICAR). During the last five years, he has been engaged in different research projects including All India Network Project on JAF. He got Young Scientist Award (YSW) from advance research of design in 2015. Apart from that, he has published 25 research papers, 30 technical and popular articles and 5 book chapters in high NAAS rated international and national journals. He is also life member in various Plant Pathological Societies. He has developed various molecular detection techniques for detection of virus and its vector. He has expertise in Plant Pathology and Plant Virology.

Preface

The course curricula and syllabi of undergraduate and post-graduate programme of Plant Pathology as recommended by Vth Deans committee of ICAR, New Delhi, has been implemented from 2017-18 academic session in all State Agricultural Universities. The course content of newly prescribed practical book "Detection and Diagnosis of Plant Diseases" has been developed as per ICAR syllabus for the UG and PG students. At present, there is no such type of practical book available which covers the entire syllabus. So, it was felt necessary to prepare a book which covers the topics of practical exercises of UG and PG courses.

The practical exercises given in the book covers acquaintance with laboratory equipments, culture media and their preparation, staining techniques of fungi, bacteria and nematodes, methods of disinfection and sterilization, isolation and purification of fungi and bacteria from soil and plant samples, *in vitro* evaluation of antagonism, mass multiplication of bio-agents and evaluation bio-efficacy of antagonists against important pathogens and methods of biochemical analyses, etc., and record the observations as per guidelines given in exercises. It is hoped that the book will prove useful for providing theory and practical training to under graduate and post graduate students.

We trust that this book will serve as the useful resource book and guide to all concerned in Agriculture, Horticulture, Forestry and in resolving plant disease diagnosis, disease etiology and disease management. We our deep sense of gratitude for the authorities of our institution for encouragement received while preparing and publishing this book.

Authors

Contents

Commonaly Used Abbreviation And Symbols

- A - Adenine or Adenosin
- ADP-Adenosine Diphosphate
- ATP- Adenosine Triphosphate
- β- Beta
- BCIP- 5-bromo -4chloro-3-indolyl phosphate
- Bp- base pairs
- BSA-bovine serum albumin
- C-cytosine or cytidine
- cDNA-complementary DNA
- CIAP-alkaline phosphatase, calf intestinal
- Cpm- count per minute
- CTP-cytosine triphosphate
- DNA- Deoxyribose nucleic acid
- DNase- Deoxyribose nuclease
- DTT-Dithiothreitol
- EDTA- thylenediamine tetra acetic acid
- EtBr- ethidium bromide
- EtOH- Ethanol
- Fmol- femto mol
- γ -gamma
- G-guanine or guanosine
- GTP-guanosine triphosphate
- HCL- Hydrochloric acid
- IPTG- β –D- isopropyl-thiogalactopyranoside
- KCL-Potassium Chloride
- Kb- Kilobase
- kDa-kilodalton
- λ-lambda
- M-moral
- mRNA-messenger RNA
- MW-molecular weight
- NaOAc- sodium acetate
- NBT-Nitro blue tetrazolium
- PAGE-Polyacrylamide gel electrophoresis
- RNA-Ribonucleic acid
- RNase-Ribonuclease
- SDS-Sodium dodecyl sulfate
- T-Thymine or thymidine
- TE-Tris-EDTA buffer
- U-Uracil or uridine
- X-Gal-5-bromo-4-chloro-3-indolyl-β-galactopyranoside

1 INSTRUMENTS USED IN THE STUDY OF PLANT DISEASES/PATHOGENS

For isolation, identification, maintenance of culture and nutritional studies of microorganism's viz., bacteria, fungi, nematodes, actinomycetes, etc., certain equipments are used as given below.

1. Autoclave

The major components of this equipment are metallic water containing vessel fitted with heating elements, safety valve, pressure gauge, etc. The equipment is used for sterilizing, culture media, water and any other heat resistant material required for microbiological studies. Moist heat i.e. steam is used for sterilization process. Steam developed in autoclave kills microorganisms by denaturing their proteins. Generally, autoclave is operated at 15 lb per sq. inch (p.s.i.) which gives rise to a temperature, of 121.6°C for 15 minutes.

2. Hot Air Oven

The major components of this equipment are a steel body containing two to three racks, heating element embedded in glass wool and a thermostat. It is mainly used for sterilization of glass wares, viz., Petri dishes, culture tubes, pipettes, etc. Dry heat at 160 - 180°C is applied for sterilizing the glass wares. Commonly time temperature relations for sterilization is 60 minutes at 160°C and 40 minutes at 170°C.

3. Laminar Flow

The major components of this equipment are working place, ultra violet light, pre-filters hepa filters, and transparent & thick plastic lid. This equipment is essential for isolation and maintenance of cultures. Before starting inoculation work, the UV light is kept on for 20-25 minutes for sterilization of inside surface of the chambers. While working, there is provision for passing sterilized air through work surface. For this purpose, two types of filters i.e. Pre-filters which removes coarse particles and hepa filters for removing fine dust particles and microbiological contaminants are used.

4. B.O.D. Incubator

The major components of this equipment are electronically controlled thermostat for maintenance of temperature, compressor and inside steel or aluminium body chamber containing three to four racks for keeping Petri dishes, culture tubes, flasks containing media inoculated the microorganisms. Generally, there is provision of a temperature range of 5 to 50°C. For routine work the temperature is maintained at 25 to 30°C, optimum temperature is 28±2°C. This equipment is essential for isolation, maintenance of microorganisms and also for conducting research on microbial nutrition, antibiotic assay, etc.

5. Microscope

Microscopes are most commonly used for observing the microorganisms which cannot be seen by naked eye. For microbiological observation a variety of light microscopes, i.e. bright field, dark field, phase contrast and fluorescent microscopes are used.

6. Centrifuge

This equipment is used for biochemical studies viz., separation of particles dispersed in a suspended matter, separation of mixtures of liquids varying in their density or solids or concentrating microorganisms in various samples for enzymatic studies, assay of various enzymes and organic chemicals like phenols. Different categories of centrifuge like, table top centrifuge, refrigerated centrifuge and ultra-centrifuge are used in the laboratory. For detection and purification of viruses from infected plants, ultra centrifuge is used.

7. Refrigerator

This equipment is a basic need for microbiological studies. This equipment is required for storing stock cultures of microorganisms at 4°C. Various antibiotics and heat sensitive chemicals are also stored in refrigerator.

8. pH meter

This equipment is used to determine the pH of nutrient media used for isolation and maintenance of microorganisms. Certain microbes are well developed at a particular pH.

9. Electronic balance

It is used for weighing chemicals for preparation of culture media and other.

10. Fermenter

The main function of a fermenter is to provide a controlled environment for the growth of microorganisms to obtain a desire product. Fermenter (bioreactor) is typically vertical, essentially closed, cylindrical steel vessels in which aerobic and anaerobic fermentations can be carried out either in batch culture or in continuous culture. This instrument is equipped to add nutrients and cultures, to remove product under aseptic conditions, and to aerate, stir and cool the mixture automatically. Biosensors and infrared monitoring can provide real-time information on the course of fermentation. Stirred fermenters can range in size from 3 to 4 liters or larger, depending on production requirements.

11. Water Bath

A constant temperature water bath is an insulated metal box fitted with temperature at the desired level. There are racks for holding test tubes or flasks. The bath is filled with water until it rises almost to the top of the racks. Water baths are usually used for melting of media, testing enzymatic activities of various microorganisms.

12. Ultraviolet-Visible Spectroscopy

Ultraviolet-visible spectroscopy or ultraviolet-visible spectrophotometer (UV-Vis or UV/Vis) refers to absorption spectroscopy in the ultraviolet-visible spectral region. This means it uses light in the visible and adjacent (near-UV and near-infrared (NIR)) ranges. The absorption in the visible range directly affects the perceived color of the chemicals involved. In this region of the electromagnetic spectrum, molecules undergo electronic transitions. This technique is complementary to fluorescence spectroscopy, in that fluorescence deals with transitions from the excited state to the ground state, while absorption measures transitions from the ground state to the excited state. UV/Vis spectroscopy is routinely used in the quantitative determination of solutions of transition metal ions and highly conjugated organic compounds. Solutions of transition metal ions can be colored because electrons within the metal atoms can be excited from one electronic state to another. The colour of metal ion solutions is strongly affected by the presence of other species, such as certain anions or ligands. For instance, the colour of a dilute solution of copper sulfate is a very light blue; adding ammonia intensifies the colour and changes the wavelength of maximum absorption (λmax). Organic compounds, especially those with a high degree of conjugation, also absorb light in the UV or visible regions of the electromagnetic spectrum. The solvents for these determinations are often water for water soluble compounds, or ethanol for organic-soluble compounds. (Organic solvents may have significant UV absorption; not all solvents are suitable for use in UV spectroscopy. Ethanol absorbs very weakly at most wavelengths.) Solvent polarity and pH can affect the absorption spectrum of an organic compound. Tyrosine, for example, increases in absorption maxima and molar extinction coefficient when pH increases from 6 to 13 or when solvent polarity decreases. While charge transfer complexes also give rise to colours, the colours are often too intense to be used for quantitative measurement.

The Beer-Lambert law states that the absorbance of a solution is directly proportional to the concentration of the absorbing species in the solution and the path length. Thus, for a fixed path length, UV/Vis spectroscopy can be used to determine the concentration of the absorber in a solution. It is necessary to know how quickly the absorbance changes with concentration. This can be taken from references (tables of molar extinction coefficients), or more accurately, determined from a calibration curve.

13. PCR Thermal Cycler

The thermal cycler (also known as a Thermocycler, PCR Machine or DNA Amplifier) is a laboratory apparatus used to amplify segments of DNA via the polymerase chain reaction (PCR) process. The device has a thermal block with holes where tubes holding the PCR reaction mixtures can be inserted. The cycler then raises and lowers the temperature of the block in discrete, pre-programmed steps.

Modern thermal cyclers are equipped with a heated lid, a heated plate that presses against the lids of the reaction tubes. This prevents condensation of water from the reaction mixtures on the insides of the lids and makes it unnecessary to use PCR oil to cover the reaction mixture. Some thermal cyclers are equipped with multiple blocks allowing several different PCR reactions to be carried out simultaneously. Also some apparatus have a gradient function, which allows different temperatures in different parts of the block. This is particularly useful when testing suitable annealing temperatures for primers.

14. Gel Doc System

Gel Doc, also known as Gel Documentation System, Gel Image System or Gel Imager, is widely used in molecular biology laboratories for the imaging and documentation of nucleic acid and protein polyacrylamide or agarose gels typically stained with ethidium bromide or other fluorophores such as SYBR Green. Generally, a Gel Doc is composed of an ultraviolet (UV) light transilluminator, a hood to shield external light sources and a camera for image capturing.

2 METHODS TO PROVE KOCH'S POSTULATES

(Pathogenicity Test)

Robert Koch (1884-1890) forwarded four essential procedural steps called postulates for correct diagnosis of a disease and its actual causal agent. These postulates are:

1. Recognition: The pathogens must be found associated with the disease in the diseased plant. The symptom of the disease should be recorded.

2. Isolation: The pathogen should be isolated, grown in pure culture in artificial media. The cultural characteristics of the pathogen should be noted.

3. Inoculation: The pathogen of pure culture must be inoculated on healthy plant of same species/variety. It must be able to reproduce disease symptoms on the inoculated plant identical to step 1.

4. Re-isolation: The pathogen must be isolated from the inoculated plant in culture media. Its cultural characteristics should be similar to those noted in step 2 (This step was added by E.F. Smith).

If all the postulates are proved true, then the isolated pathogen is identified as the actual causal organism responsible for the disease.

3 PREPARATION OF CULTURE MEDIA FOR THE GROWTH AND ISOLATION OF FUNGI AND BACTERIA

Media is the substrate used for isolation and culture of micro-organisms such as bacteria, fungi, actinomycetes, etc., in the laboratory. Different types of media are used for culture and growth of micro-organisms under laboratory conditions as given below:

Culture: The active growth of the micro-organism obtained in pure form in laboratory is known as culture.

Culture medium or substratum: Any material which supplies the nutrients and supports the growth of the micro-organism is known as culture medium. In layman's language it may be called as food for the microorganisms.

Contamination: Entry of any other organism other than desired in the tube, flask or Petri plate containing pure culture and growing side by side with the desired organism is known as contamination.

Pure culture: Culture contains organism of one type only then it called pure culture.

Requirement of a good culture medium

(1) It should have the proper nutrients in the proper proportions.

(2) It should have proper pH.

(3) It should be sterile i.e. free from all contaminating organism.

Nutrient for a culture media

The main elements required for bacterial growth are carbon and Nitrogen. Water supplies the hydrogen and sometime oxygen and also acts as the carrier for the different nutrients. Carbon is usually supplied in the form of carbohydrates and nitrogen may be in the organic form. In the inorganic form it is given in the form of citrates or salt of Ammonia. In the organic form is given in the form of extract of animal tissue or extract of vegetables. In the laboratory, peptone is most commonly employed form of organic nitrogen also supplied carbon and minute amounts of phosphorous. Potash, sulphur and magnesium are also necessary for bacterial growth.

Culture media are either used in the liquid state or after solidifying then by the addition of some gel like substance. Solid media are useful in studying the growth characters of bacteria and for obtaining pure culture by sub culturing,etc. Liquid media are generally used to grow bacteria on large scales and to study their biochemical reaction.

Culture media are divided into two categories on the basis of their composition.

(1) Synthetic media: Media whose exact chemical composition is known are termed as synthetic media, with the help of such medium, nutritional requirements of the organism can be determined accurately e.g. Rechards solution (Czapek- dox medium).

(2) Non synthetic media: Media composed of ingredient of unknown chemical composition are known as non-synthetic media. Some of these ingredients are peptone, before meat extract, blood serum, agar, potato, gelatin etc. These media are needed for organism which do not grow on synthetic media as they requires certain type of vitamins, hormones or enzymes which are not present in non-synthetic media. Examples are Nutrient broth, nutrient agar, potato dextrose agar (P.D.A.), etc.

The non-synthetic media can be further classified into two groups.

(i) Natural Media: Natural medium is of unknown chemical composition and is prepared from substance gathered form natural sources e.g. Slice of potato, vegetable and fruits, dung, host decoction, etc.

(ii) Semi synthetic media: Semi synthetic media are those which are made up of organic and inorganic ingredients. Hence, their chemical composition is partially known e.g. potato dextrose agar. Any media which contains agar comes in either natural or semi synthetic categories.

Synthetic and non-synthetic media are further classified into four groups on the basis of their use.

(i) General purpose medium

(ii) Selective medium

(iii) Differential medium

(iv) Special medium

(i) General purpose medium: A large number of micro-organism can be grown on these medium and therefore they are used for routine isolation and maintenance of microbes. e.g. PDA, nutrient agar, etc.

(ii) Selective medium: Media which favor the growth of particular type of organism and prevent the growth of others are known as selective media. These types of media help in selecting desired organism in pure form. e.g. - Yeast extract mannitol agar (YEMA) for the Rhizobium, Trichoderma selective medium

(iii) Differential media : Media which contain substances used for a specific purpose as those that contain a carbohydrate and an indicator, making it possible to distinguish organism that ferment the carbohydrate from those that do not other different media contain acetate, gelatin, etc.

Preparation of culture media:

(a) Nutrient Broth :

Chemical composition	
Meat or beef extract	: 3 g
Peptone .	: 5 g
Distilled water	: 1000 ml
NaCl	: 3 gm

Procedure:

Dissolve meat extract in water and mix peptone in it. Boil the mixture till peptone is dissolved. Quantity of the final product will be reduced due to evaporation. Makeup the volume by adding more water. Filter through cotton and adjust the pH 7.0. Fill the test tube or flasks, plug them and sterilize in autoclave at 15 p.s.i. for 20 min.

Use- It is generally used for growth and storage of microorganism specially bacteria.

(b) Nutrient agar: Dissolve 20 g agar-agar in one liter nutrient broth, resulting media will be nutrient agar. It is also general purpose medium useful for growth of bacteria.

(c) Potato Dextrose Agar (PDA)

Composition	
Pealed potato	200 g
Dextrose	20 g
Agar	20 g
Distilled water	1000 ml

Procedures:

Weight 200 g of previously peeled potatoes. Cut it into small pieces. Boil these pieces with water (approx. 500 ml) till the pieces of potatoes become soft. This can be checked by pressing the piece in between finger and thumb. Filter the boiled potatoes with the help of muslin cloth. Discard the pieces and collect the decant for further use. Add 20 g of dextrose in the same and mixed it thoroughly. Similarly, agar-agar 20 g may be dissolved in the separate 200 ml flask filled with lukewarm distilled water. This may be heated to dissolve all the flacks of agar or entire agar powder. Mix both the content and makeup the volume 1000 ml. Fill in the test tubes or flask, plug and sterilize at 15 p.s.i. for 20 min in the autoclave. It is also general purpose medium for the cultivation of fungi.

(d) Eosine Methylene blue agar:

Composition	
Peptone	10 g
K_2HPO_4	2 g
Agar	20 g
Distilled water	1000 ml

Mix all the ingredients with the 1000 ml distilled water. Bring to the boil and continue to heat until the agar is dissolved. Make up the volume lost due to evaporation. No adjustment of the pH is necessary. Dispense 100ml into flasks. Sterilize in an autoclave at 15 p.s.i. for 20 min.

It is a blend of two stains, eosine and methylene blue in the ratio of 6:1.

When ready to use, melt a flask of agar (1000 ml) in water bath and add the following.

Lactose	1 g
Eosine (2%) solution	1.25 ml
Methylene blue(0.5% sol)	1.25 ml

Mix the contents of the flasks thoroughly steam in an Arnold's sterilizer for 5 min. Allow the agar to cool to about 50°C then pour into sterile Petri plates. This is slightly selective media for some Gram-negative bacteria which slightly inhibits the growth of Gram-positive bacteria and provides a color indicator distinguishing between organisms that ferment lactose (e.g., *E. coli*) and those that do not (e.g., *Salmonella, Shigella*). Organisms that ferment lactose display "nucleated colonies" colonies with dark centers.

(e) Czapek-dox Agar medium

$NaNO_3$	2.0 g
K_2HPO_4	1.0 g
$MgPO_4.7H_2O$	0.5 g
KCl	0.5 g
$FeSO_4$	0.01 g
Sucrose	30.0 g
Agar agar	20.0 g
Distilled water	1000 ml

(f) Richards Media

Potassium nitrate	10.0 g
Potassium dihydrogen phosphate	5.0 g
Magnesium sulphate	2.5 g
Ferric chloride	0.02 g
Sucrose	5.0 g
Agar	15.0 g
Distilled water	1000 ml
pH	6.6-7.2

Dehydrated media: Media from which water has been removed are called dehydrated media. They are also form generally on a commercial scale. They are also known as ready to use media.

Advantages of dehydrated media:

(1) They are relatively stable.

(2) The preparation of media is a time saving and economic process.

(3) Following dehydration, addition of distilled water required no filtration.

(4) Adjustment of the pH of the medium not required.

The compositions of each medium and final pH of the sterilized medium at 25°C are stated on the table of the bottle. All that is necessary to dehydrate the medium is to weight accurately the contents of dehydrated medium, dissolve the power is fresh distilled water and sterilize the solution.

The pH of culture media

Culture media are adjusted to different degree of acidity or alkalinity. In biology it is usually expressed in terms of negative hydrogen ion concentration (pH). The term pH is defined as the logarithm of the reciprocal of the hydrogen ion concentration. For each kind of micro-organism there is a change of reaction of media in which good growth will take place. Some organism grow best in acidic medium, still other prefers media that are approximately neutral in pH. Most organisms grow best in culture media adjusted to a pH of about 7.0. It is therefore necessary to adjust the pH of the prepared media to the organism to be studied unfavorable influence of many salts included in fluid having an acidic pH. Coagulation by heat occurs more rapidly in acid solution. Thus milk which is only slightly sour may odor on being warmed.

Buffers: Substances that possess resistance to the changes in acidity or alkalinity are known as buffers. The commonly used buffers are salts of weak acids. These include phosphate carbonates etc. In addition amphoteric components such as proteins and protein split products (proteases, peptones, peptides, amino acids) are also efficient buffers.

Adjustment of pH: In order to adjust the reaction to desired pH of 7(Neutral), 0.1N NaOH or 0.1N HCL is added drop by drop until a desire pH is achieved. Bromothymol is used in the media to observe the changes in the pH. This has yellow colour in acidic media, blue in alkaline and grass green colour in neutral media.

Use of ingredients:

Water: Water is absolutely necessary for the existence of living cells. Tap water may show considerable variation from one locality to another with the result that uniform culture medium cannot always be prepare. Since distilled water is of definite composition, it should be used in preference tap water for the preparation of culture media.

Peptone: Peptone is intermediate products of hydrolysis form by the action of certain protolytic enzyme on natural proteins. The most important function of peptone in culture media is to furnish available source of nitrogen. Since the amino acids are amphoteric compounds, peptones are also excellent buffers.

Beef or meat extract: Beef extract is added to media to supply certain substances that stimulate bacterial activity. It contains enzymes which cause growth of micro-organism. Beef extract also contains nitrogenous and non-nitrogenous compounds.

Gelatin: It is a protein which is not soluble in cold water, but swells and softens when increased in it. It is quite soluble in being water so on cooling it solidifies to form a transparent gel. Gelatin is rarely used as a substitute for agar for preparation of solid media because

(a) It is attached and decomposed by many bacteria.

(b) It melts at 30°C room temperature during summer days.

Agar powder or agar shreds: It is complex carbohydrate refine from marine red algae *Gelidium sp*. (primarily from the genera Gelidium and Gracilaria, or seaweed *(Sphaerococcus euchema)*. Commercially it is derived primarily from *Gelidium amansii.*) and it is useful agent for solidification of media. Agar must be capable of producing a clear solution when hot. pH of the media containing agar is increase by 1.5 to 2 % therefore, the pH of media containing agar is adjusted little low when hot.

NaCl: Sodium chloride is commonly added to culture media to increase osmotic pressure, although this usually not necessary. NaCl does not act as a buffer.

Fermentable compound: There are two functions of fermentable compounds in culture media.

(a) They furnish readily available source of energy provided. These organisms elaborate the enzymes necessary to ferment the compounds.

(b) Fermentation reactions are of great help in identification and classification of organism.

The fermentable compounds usually added to culture media includes: glucose, sucrose, cellulose, lactose, starch, galactose, dextrin, maltose, etc.

Observations:

Inoculation of the given culture of fungi to agar slants and Petri dishes containing PDA.

4 ISOLATION OF PLANT PATHOGENS

Isolation: It is a process by which separate the causal organism from the diseased tissue of the host and obtaining it into pure form in culture medium or the process of separating micro-organism from the diseased/infected tissue of host.

Isolation of pathogen is done to

- Get organism in pure form to determine the host pathogen relationship.
- To study the beneficial or harmful effect of particular microorganism.
- To prove the Koch's postulates (proving the pathogenicity).
- To study its life cycle, cultural characteristics and physiology of pathogen.
- For identification of plant pathogen.

Pure culture:-

- The growth of single species of microorganism on a medium is called pure culture.
- A pure culture is required for studying the morphology and physiology of microbes.
- R. Koch (1875) was the first scientist to introduce the pure culture technique in microbiology.

Mixed culture:- It is growth of two or more species of microbes in the culture medium.

Plate culture:- It consists of an organism growing on solid medium contained in a Petri plate.

A slant culture:- It consists of an organism on the inclined surface of solid medium in a test tube. This is also referred to as nutrient agar slant culture.

Liquid / Broth culture:- The liquid medium such as nutrient broth, milk or dextrose broth containing the growth of an organism is called broth culture.

Stab culture:- It is prepared by stabbing a solid medium (N.A.) to a considerably depth with sterilized wire needle such as nutrient gelatin or nutrient agar.

Methods of pure culture

1. The spread plate method.
2. The pour plate method.
3. The streak plate method.

1. Spread plate method:- Serial Dilution Method

Requirements

- Sample
- Six sterile Petri plates containing specific media.
- Six test tubes containing 9 ml sterile distilled water
- Glass spreader
- Nutrient agar slants
- Pipette

Procedure:

- Label all the Petri plates and test tubes form one to six.
- Take one gram of soil or other sample from the bacteria need to isolated and mix in 10 ml sterile water. This will give 10^1 dilution.
- Take 9 ml water in remaining all the tubes
- Add 1 ml solution (soil : water) from the first tube to the second. This will give 10^2 dilution.
- Now the second tube will have 10 ml of solution and the same.
- From the tube (second tube) take 1 ml solution (soil : water) and mix in the third tube. This will give 10^3 dilution.
- The same should be repeated till the last test tube (six tubes). This will give 10^6 dilution.
- Take 0.1 ml of the solution from the each tube and spread the help of spreader on the different labeled Petri plate containing specific media.
- Incubate the plate at suitable temperature
- Observe the plate after 24 hours
- Individual isolated colony represent the cfu (colony forming unit)
- Individual colony can be transfer to some other plate or slant for the pure culture
- Colonies can be counted on the colony counter and total number of the microorganism can be calculated by the formula presented hereunder.

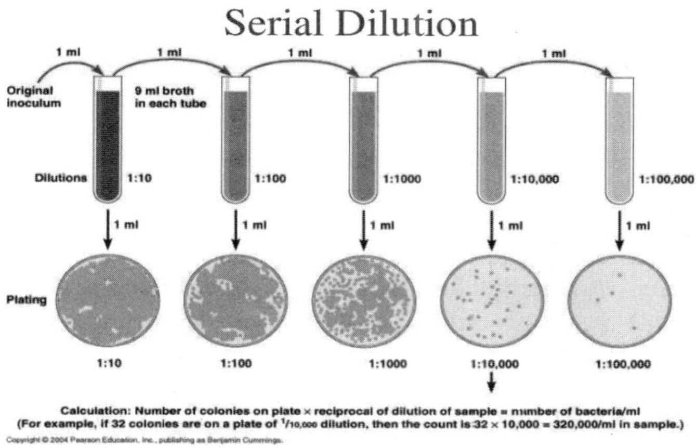

Formula:- No. of cells/ ml or g of sample (Cfu/ ml)

Cfu/ ml = Average no. of colonies × Dilution × Volume taken

E.g. one gets 30 colonies in 10^5 dilutions.

Then no. of cfu/ ml = $30 \times 10^5 \times 0.1 = 30 \times 10^5 \times 10 = 30 \times 10^6$

Pour plate method:-

The principle of this method is to dilute the inoculum in successive tubes containing liquefied agar medium and throughout distribution of cells within the medium. The inoculum then poured into Petri dishes and incubated to get well isolated colonies.

- Mixed broth culture of bacteria.
- Three sterile Petri dishes.
- Three tubes of nutrient agar medium.
- Three tubes for nutrient broth.
- Water bath and thermometer.
- Four nutrient agar slants.

Procedure:-

- Dilute the culture several times by transferring 1 ml of culture to 9 ml of sterile distilled water tubes as shown in serial dilution method.
- Melt 2-3 tubes containing nutrient agar; put them at 45°C in water bath.
- Allow the melted medium to cool.
- Inoculate 1 ml of highest dilution in to cool melted agar.
- Mix contents of the tube well by rotating the tube between palms of the hand.

- Pour the inoculated nutrient agar tube in to sterile Petri dish.
- Similarly second dilution is also poured into sterile Petri dish.
- Allow the agar to solidify at room temperature.
- Mark the plate, invert and incubate at 37°C for 24 hours.
- Proceed further in the manner similar to the streak plate method.

Streak plate method:-

It is also called as the rapid qualitative isolation method. It is essentially a dilution technique that involves streaking a loopful of culture over the surface of an agar plate. Although many types of procedure are performed the four ways of quadrant streak is commonly adopted.

Procedure:-

- Take loopful of a mixed culture and make streak on the surface of nutrient agar medium in the Petri dish.
- Similarly with the same loop, after flaming it without taking culture again rotate the plate streak with the help of wire loop.
- Again rotate the plate after flaming the wire loop streak on the surface of nutrient agar medium on the Petri dish.
- Flame the wire loop. Rotate the plate again at 45° angle again streak for the fourth time on the surface of nutrient agar on Petri plate.
- Invert the Petri dish and incubate it at 37°C temp. for 24 hrs to 36 hrs.
- Observe and record the observation.
- Transfer the well separated colony on nutrient agar slant from the Petri dish to get pure culture. Plates showing growth of bacterial cells after streaking

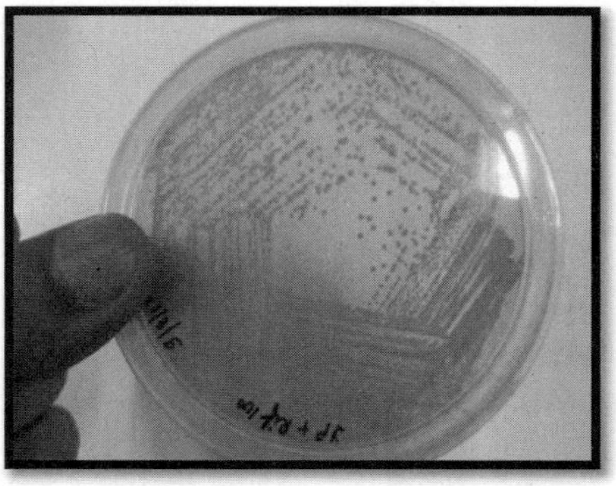

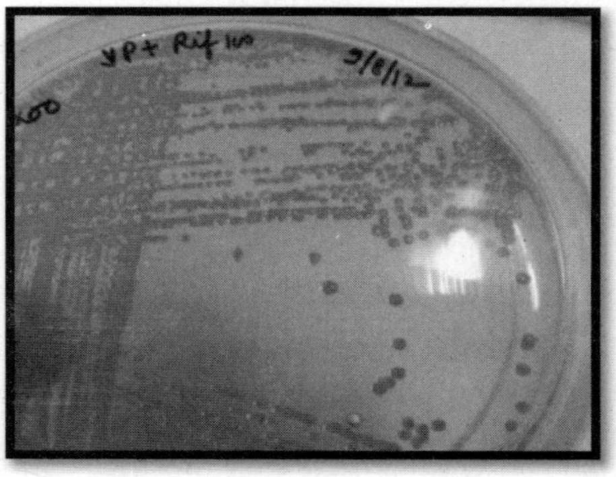

1 Isolation of the fungi from the infected/diseased host

- This method is also called as tissue method and is followed for the isolation of pathogen particularly fungal pathogen from the host tissue when the pathogen is situated deep in the tissue.
- In this method, pieces of host infected tissue are transplanted/placed on sterilized agar plates or test tube slants for activating the fungal growth or diseased tissue may be kept in moist chamber.

Procedure:

- ➢ Select a tissue from the most advanced portion of the disease part.
- ➢ Cut the disease piece from the infected leaf of square or rectangular shape.
- ➢ Cut the infected portion with a blade in such a way that each piece should have a part of infected/diseases/necrotic/infected part showing symptoms and three parts showing healthy part.
- ➢ In order to eliminate the contaminants from the outside of the host surface disinfection is necessary. It is carried out by 0.1% mercuric chloride solution for 30 sec to 1 min or 2.5% sodium hypochlorite solution for 2-3 min. This should be done in the controlled condition so that contaminants cannot land on the tissue after disinfection.
- ➢ Wash with 3-4 times with sterilized distilled water after surface disinfection to remove excess of disinfectant.
- ➢ Transfer of diseased piece on media (Potato dextrose agar medium) previously filled Petri plates or slant for the active growth of the tissue.
- ➢ Place the plate at appropriate place in incubator after proper labeling.

Process:

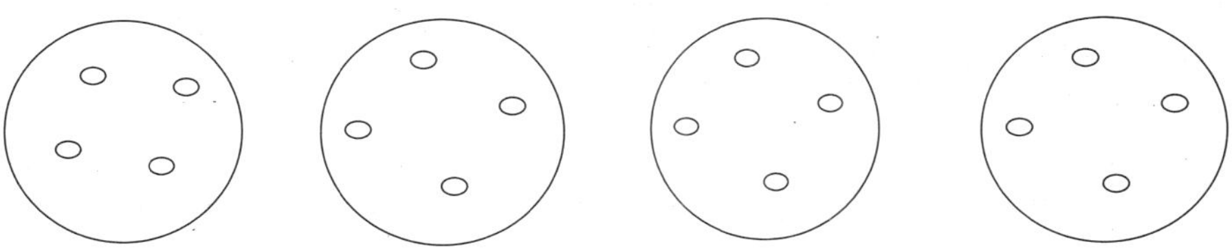

2 Isolation of bacteria from the infected tissue

Tissue isolating is not carried out for the bacterial disease as bacteria survive inside the cell. Bacteria do not have positive movement. However, when the bacterial disease infected tissue is cut in the presence of water, many bacterial cells come out of the cell due to rupture of the cells in the form of cloud. This can be observed very easily under the microscope. This cloud of the bacteria cell is known as 'ooze'. If the cell is severely infected and quite large size of the piece is cut and dipped in the water the ooze can be seen in the water without the aid of microscope.

Procedure:

Diseased samples & infected part of leaves, stems, roots etc. → Cut the samples with the help of blade/knife → Rins the samples with 0.1% HgCl2 solution → Rins the Samples with sterilized distilled water for three times → Pouring media (PDA) in sterilized plates → Put the sample bits (four) in each plates.

- Select diseased tissues.
- Surface dis-infectant i.e. 0.1% mercuric chloride or 2.5% sodium hypochloride solution.
- Washing with distilled sterilized water.
- Crush the tissue in the small plate; tube etc. with the help of sterile mortar or any other sterile object.
- These crushed tissues can be streak (line) with the help of wire loop on a Petri plate, filled with solidified Nutrient Agar media.
- To streak, a loop is first dip in the crushed tissue and then aseptically touches on the surface of media in the plate and loop is gradually moved on Petri plate in such a way that it will make zigzag line on the plate. Care should be taken to avoid piercing the media with the loop while making streak. This is known as streak plate method.
- Plates are incubated in inverted position for 24 hours at approximately 37°C temperature in incubator.

Streaking | Bacterial Colony | Single colony | Pure culture in slants

3 Isolation of bacteria from the Soil

Microorganism including fungi, bacteria, vascular arbascular mycorrhiza, etc can be isolated from the soil by serial dilution method. In serial dilution, one gram soil is taken and is mixed in 10 ml sterile water. Similarly, 1 ml of the solution from first tube is taken and dispensed/added in another tube filled with 9 ml of sterilize distilled water. This will make total volume of the tube 10 ml and the solution will be 1:100 time diluted. Similarly, if the solution is diluted six times, this will give 10^{-6} fold dilution of the sample. Hundred micro-liters to 1 ml of the solution is taken and spread on the Petri plate filled with the desired media. This is known as **spread plate method**. Plates are incubated in inverted position for 24 hours at approximately 37°C temperature in incubator. This will dilute the microorganisms and will give very clear colonies of the bacteria at sufficient distance to further transfer in another tube or plate for the purification. Even if the colonies are appear in the media very close to each other making isolation difficult, these may further be diluted to 10^{-7} - 10^{-8} fold.

Diseased samples (Infected part of leaves, stems, roots etc.) → Cut the samples with the help of blade/knife → Rins the samples with 0.1% $HgCl_2$ solution

→ Streak the crushed samples solution or put the samples bits (four) in each plate. → Streak the crushed samples solution or put the samples bits (four) in each plate. → Streak the crushed samples solution or put the samples bits (four) in each plate.

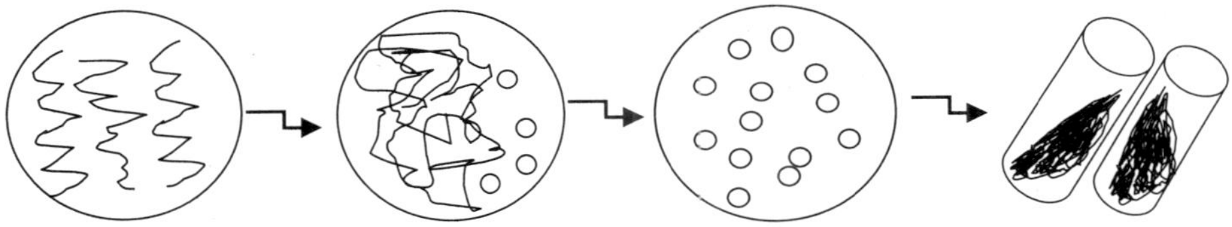

5 STAINING OF FUNGI

Staining of fungi is done to make their microscopic details more clearly visible. The use of a single stain to colour an organism is commonly referred to as simple staining. The differential stains are used for making visible specific details such as the nucleus in the fungi and the flagella

Requirement:

Microscope, Glass slide, Cover slips, Needle, Blotter paper, Viable sporulating culture of fungus, Cotton blue in lactophenol.

Cotton blue

Saturated solution of cotton blue in 95% alcohol - 10.00 ml
Glycerin - 10.00 ml
Water - 80.00 ml

Cotton blue in lactophenol:

Phenol (Crystals)	-	20.0 g
Lactic acid	-	20.0 g
Glycerol	-	40.0 g
Water	-	20.0 ml

Melt phenol in water until it is dissolved and then add lactic acid and glycerol. Add 0.05 g of cotton blue per 100 ml

Procedure:

- Place a drop of lactophenol cotton blue on clean glass slide
- Transfer small amount of fungus with spores and sporulating structure into the drop with the help of sterile needle Transfer small amount of fungus with spores and sporulating structure into the drop with the help of sterile needle
- Gently tease the material fungal structure and mix with stain
- Place a clean cover glass slip over the preparation taking care that air bubbles are removed
- Remove air bubbles from preparation by gentle heating and pressing
- Remove the excess stain with clean filter paper/blotter paper. The prepared slide is ready for microscopic observation
- To make the slide semi-permanent, apply a thin layer of nail-polish around the edge of cover glass slip. Allow the preparation to dry over-night and apply a second coat of polish

- Observe the slide under low/high power in microscope or in oil immersion and draw the fungal structure and identified

Observations:

Prepare the given fungal slides and draw the diagrams of various structures of fungi

6 STAINING OF BACTERIA

Living bacteria are colourless and transparent and therefore their detailed structures cannot be observed unless a phase-contrast microscope is used. This difficulty however can be overcome by the use of selective dyes which make them more contrast in colour with their surroundings staining, therefore, is of primary importance for the recognition of bacteria.

The stains are coloured organic compounds (dyes). Based on their chemical behavior, the dyes are classified as acid, basic and neutral. An acid (anionic) dye has a negative charge and basic (cationic) dye carries a positive charge. A neutral dye is a complex salt of a dye acid with a dye base. Acid dyes stain basic cell components and basic dyes stain acidic cell components.

In a bacterial cell some of its parts are acidic in reaction while other parts are basic. The acidic constituents of the cells such as nuclei react with the basic stains and the basic constituents of the cells such as cytoplasm react with acid stains. Bacterial cells are slightly negatively charged when pH is near neutral. A stain such as methylene blue has a positively charged blue portion of the molecule that stains the microorganism. Nigrosin and India ink are frequently use for negative staining of microbial cells and this type of staining is particularly useful for viewing some structures such as capsules that surround some bacterial cells.

Simple stains:

Simple staining is to study the morphology and arrangement of bacterial cells. The most commonly used basic dyes are methylene blue, crystal violet and carbol fuchsin. The exposure time differs for each stain. Carbol fuchsin requires 15-30 sec; crystal violet 20-60 sec; methylene blue 1-2 min. These stains can be easily prepared, are less subjected to deterioration, give rapid result with one application.

Simple stains can be divided into two categories:

(i) Basic dye

(ii) Acid dye.

Bacterial cells are composed of acidic constituents and therefore, basic dyes like methylene blue, crystal violet and basic fuchsin, eosio- safranin -0, etc; are widely used for staining. Generally the entire cell is uniformly stained. Some acidic dyes, like sodium eosine and nigrosine are also used to study bacterial staining. Being acid dyes they will no combine with the acid constituents of bacterial cell, rather such dyes will deposit around the bacterial cells, leaving the bacterial cells colorless, indicating the shape and size of the bacteria. This is also called negative staining.

A - Simple staining:

Simple staining of bacteria with basic dye

For routine observation of any bacteria we can use any of the following two basic dyes:

Leoffler's methylene blue

Saturated solution of methylene blue in a alcohol	300 ml
KOH (0.01%) in water	1000 ml

Stain for 3 min; then wash with distilled water.

Crystal violet solution

Crystal violet	10 g
Absolute alcohol (100% ethanol)	100 ml
Distilled water	900 ml

Dissolve the dye in the alcohol, filter and add the water

Procedure

Prepare bacterial smear on glass slide. For this, take a loopful of bacterial suspension on a clean glass slide and spread it using inoculating needle and allow the slide to dry. Add few drops of the stain on the smear and allow the stain to remain for one minute. Wash the stain in tap water, drain, blot and dry in air or hold high over a flame. Examine the slide under the oil-immersion objective of the microscope.

Observations - Record morphology of bacteria

B. Differential staining of Bacteria:

In 1884, a Danish physician named, Christian Gram discovered a new technique to differentiate the bacteria of similar morphology. He used two dyes viz., crystal violet and safranin in sequence each of a different colour. The organisms that retain the colour of the first dye i.e. crystal violet is called Gram- positive and those that cannot retain the dye when washed with a decolorizing solution and take the colour of the second dye are called gram- negative. This is known as Gram staining.

Principle of Gram staining

The cells walls of Gram-negative bacteria are thinner than that of Gram-positive bacteria and contain higher percentage of lipid content. During the staining of Gram negative bacteria, the alcohol treatment extracts the lipid. This results in increased porosity or permeability of the cell. The crystal violet-iodine complex, thus can be extracted and the Gram-negative bacteria is decolourized. The cells subsequently take up the colour of the counter stain safranin.

The cell wall of Gram- positive bacteria with lower lipid content becomes dehydrated during alcohol treatment. The pore size decreased, permeability is reduced and the crystal violet-iodine complex cannot be extracted. Therefore, the Gram-positive cells remain purple- violet.

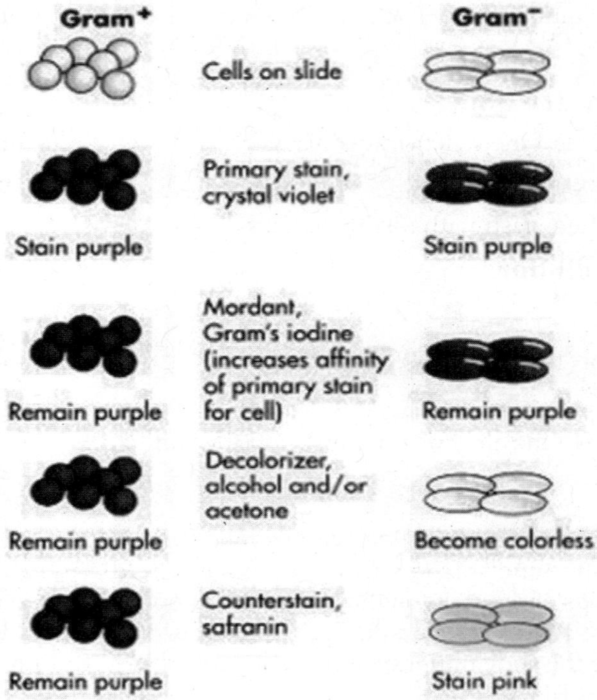

Comparison of Gram-positive and Gram-negative cells

Preparation of staining reagents: (for Hacker's method of gram staining)

Primary stain-

Sol. A:	Crystal violet (certified 90 dry content)	2 g
	Alcohol 95% (vol/vol)	20 ml
Sol. B:	Ammonium oxalate	0.8 g
	Distilled water	80.0 ml

Mix A and B sol obtain the crystal violet staining reagent. Store for 24 hours and filter through paper before using.

Mordant:

Iodine	1 g
Potassium iodide	2 g
Distilled water	300 ml

Grind the iodine and potassium iodide in a mortar and add water slowly with continuous grinding until the iodine is dissolved.

Decolourizing agent - Alcohol 95% (V/V)

Counter stain - Safranin 2.5% in 95% (v/v) alcohol -10 ml, Distilled water -100 ml

Procedure:

- Air dry or heat fixes the bacteria smear. Flood the slide with crystal violet staining reagent for 1 minute.
- Wash the smear in a gentle and direct stream of tap water for 2 seconds.
- Flood the slide with iodine mordant for 1 minute.
- Wash the slide in a gentle and indirect stream of tap water for 2 seconds
- Blot the smear dry with absorbent paper.
- Immerse the smear in 95% (v/v) alcohol for 30 seconds with gentle agitation.
- Blot the smear dry with absorbent paper.
- Immerse the smear for 2 minutes with counter-stain.
- Wash the smear in a gentle and indirect stream of water until no colour appear in the wash water.
- Blot the smear dry with the absorbent paper.
- Examine under the microscope

Observations- Prepare the slide of bacterial smear and differentiate from Gram-negative to Gram-positive bacterial cells.

7 ISOLATION AND PURIFICATION OF PLANT PATHOGENIC FUNGI FROM RHIZOSPHERE SOIL

Soil contains varied group of microorganism's viz., bacteria, fungi, actinomycetes, algae and protozoa. The total number and types of microorganisms present in soil vary depending on chemical composition, physical conditions, organic matter contents, agricultural practices, etc. Some of the microbes cause diseases in plants like damping - off, root rots, wilt, etc. However, a large number of microbes play an important role in decomposition of composts and plant residues and biogeochemical cycle.

Requirements: Various media like Potato Dextrose Agar, Czapek-dox and Martins' Rose Bengal Agar medium, Fusarium selective medium, Trichoderma selective medium, etc; for fungi, Sterilized Petri plates, Sterile distilled water blanks (90 ml in each 250 ml. conical flask), Soil samples.

Composition of Martins' Rose Bengal Agar medium

Dextrose	10.0 g
Peptone	5.0 g
KH_2PO_4	1.0 g
$MgSO_4$	0.5 g
Rose Bengal	0.05 g
Streptomycin	0.03 g
Agar-agar	20.0 g
Distilled water	1000 ml

Dissolve 1 g of streptomycin sulphate in 100 ml sterile distilled water, after opening the vial aseptically; add 0.3 ml of streptomycin sulphate in solution to each 100 ml of the basal medium after it is cooled. This is a general purpose medium for the isolation of fungi from soil.

Czapek-dox Agar medium

$NaNO_3$	2.0 g
K_2HPO_4	1.0 g
$MgPO_4.7H_2O$	0.5 g
KCl	0.5 g
$FeSO_4$	0.01 g
Sucrose	30.0 g
Agar-agar	20.0 g
Distilled water	1000 ml

Fusarium selective medium

Peptone	15.0 g
Dipotassium hydrogen phosphate	1.0 g
Magnesium sulphate	0.5 g
Agar-agar	15.0 g
Penta chloro nitro benzene	0.5 g
Oxgall	0.5 g
Streptomycin sulphate	0.1 g
Chlortetracycline hydrochloride	0.05 g
Distilled water	1000 ml

Trichoderma selective medium

Ammonium nitrate	1.0 g
Dipotassium hydrogen phosphate	0.9 g
Magnesium sulphate	0.2 g
Potassium chloride	0.15 g
Glucose	3.0 g
Agar-agar	15.0 g
Rose Bengal	0.15 g
Chloromphenical	0.25 g
Metalaxyl	0.30 g
Pentachloronitrobenzene	0.20 g
Distilled water	1000 ml

Procedure:

- Prepare the Petri plates containing agar media.
- Add 10 g field soil (air dry) in 90 ml distilled sterilized water in conical flask and shake it for 3 to 4 minutes so that microbes get suspended in water. This will make 10^1 dilutions.
- From 10^1 suspension, take 10 ml and transfer it to 90 ml sterilized distilled water which will make 10^2 dilutions.
- Serial dilutions are made up to 10^7 following the said procedure.
- For determining fungal population, transfer 0.2 ml suspension of 10^3, 10^4 and 10^5 dilutions to plates containing PDA and spread it uniformly using glass spreader.

- For determining bacterial population, transfer 0.2 ml suspension of 10^5, 10^6 and 10^7 dilution to Petri plates containing Nutrient Agar (NA) medium and also spread uniformly using glass spreader.
- Incubate the plates at 25°C for three days and observe the microbial growth after 24 hours onwards periodically.

Observations

- Count the fungal and bacterial colonies.
- Stain the microbes and observe under microscope.
- Determine the number colonies using following formula:
- No. of colonies (x) X 5 X dilution factor
- (Say 10^3 for fungi, 10^5 for bacteria)
- Draw the diagrams of fungi from colonies appeared in Petri plates.

8 ISOLATION AND PURIFICATION OF PLANT PATHOGENIC BACTERIA FROM RHIZOSPHERE SOIL

Soil contains varied group of microorganisms' viz., bacteria, actinomycetes, algae and protozoa. The total number and types of microorganisms present in soil vary depending on chemical composition, physical conditions, organic matter contents, agricultural practices, etc. Some of the microbes cause diseases in plants like damping - off, root rots, wilt etc. However, a large number of microbes play an important role in decomposition of composts and plant residues and biogeochemical cycle.

Requirements:

Nutrient Agar medium and Pseudomonas Agar fluorescens Media are mainly used for isolation of bacteria from rhizosphere soil.

Sterilized Petri plates, Sterile distilled water blanks (90 ml in each 250 ml. conical flask), Soil samples.

Composition of Pseudomonas Agar fluorescens medium (g/liter)

Casein enzymic hydrolysate	10.0 g
Protease peptone	10.0 g
Di-potassium hydrogen phosphate	1.5 g
Magnesium sulphate	1.5 g
Agar-agar	15.0 g
pH	7.0 ± 0.2
Distilled water	1000 ml

Method: Thirty eight gram readymade preparation was suspended in 1000 ml distilled water containing 10 ml glycerol. The mixture was then boiled and the desired quantity of media was transferred to Erlenmeyer flasks and culture tubes. The media was sterilized in autoclave at 15 p.s.i. for 15 minutes.

Procedure:

- Prepare the Petri plates containing agar media.
- Add 10g field soil (air dry) in 90 ml distilled sterilized water in conical flask and shake it for 3 to 4 minutes so that microbes get suspended in water. This will make 10^1 dilutions.

- From 10^1 suspension, take 10 ml and transfer it to 90 ml sterilized distilled water which will make 10^2 dilutions.
- Serial dilutions are made up to 10^7 following the said procedure.
- For determining fungal population, transfer 0.2 ml suspension of 10^3, 10^4 and 10^5 dilutions to plates containing PDA and spread it uniformly using glass spreader.
- For determining bacterial population, transfer 0.2 ml suspension of 10^5, 10^6 and 10^7 dilution to Petri plates containing Nutrient Agar (NA) medium and also spread uniformly using glass spreader.
- Incubate the plates at 25°C for three days and observe the microbial growth after 24 hours onwards periodically.

Observations:

- Count the fungal and bacterial count.
- Stain the microbes and observe under microscope.
- Determine the number colonies using following formula:
- No. of colonies (x) X 5 X dilution factor

  (say 10^3 for fungi, 10^5 for bacteria etc.)
- Draw the diagrams of fungi from colonies appeared in Petri plates and prepare the slide of bacterial smear and identify the bacterial reaction whether, it is Gram (+) or Gram (-).

9 EVALUATION / BIOASSAY OF FUNGICIDES - INHIBITION ZONE TECHNIQUE AND SLIDE GERMINATION TECHNIQUE

Requirements:

Petri dishes, flasks, PDA medium, autoclave, hot air oven, laminar flow, inoculation needle, spirit lamp, culture of fungal and bacterial pathogen, test fungicides, incubator.

a) Inhibition zone technique:

This technique was introduced by Thornberry (1950) and called as paper disc plate method.

Procedure:

- Prepare PDA and sterilize it.
- Pour 15 ml of sterilized hot (80 to 90°C) agar medium into the bottom of 90 mm flat-bottomed Petri plates.
- After solidification, add 5 ml of lukewarm seeded agar containing (0.5 - 1 x 10^6 spores/ml) and spread uniformly the seeded agar.
- Transfer the Petri plates to refrigerator and stack upside down to prevent condensation of moisture.
- Prepare required concentration of the fungicide using sterilized distilled water.
- Dip the sterilized filter paper discs of 5 mm diameter in the test chemical and transfer aseptically to seeded-agar plates after drying.
- Place three discs per plate (maintain at least three such plates for each chemical)
- Incubate the inoculated Petri plates at 28 ± 2°C for four days.
- Observe for clear zones of inhibition around each disc and measure from the centre of the filter paper disc.

Note: From these observations, the minimum inhibition concentration (MIC) is recorded. The MIC is the dosage at which no growth occurs on the impregnated seeded disc over a period of 7-10 days. It serves as a basis for comparing the efficacy of different test chemicals.

b) Slide germination technique

Aim: To test the effect of fungicides on spore germination slide germination technique was introduced by Reddick and Wallace (1910) to determine fungicidal value of spray mixture or solution.

Materials required: Clean glass slides (3 x 1" size), spore suspension of test fungus, Petri plates with three layers of blotting papers, haemocytometer, microscope, pipettes, test chemicals (fungicides), sterilized distilled water, beakers, etc.

Procedure

- Prepare spore suspension of test fungus grown on PDA (7 days old) by using sterilized distilled water and adjust the final concentration to 50000 spores/ml using Haemocytometer.
- Prepare required concentrations of test fungicide using sterile distilled water.
- Place one drop of test fungicide at the center of clean and sterilized glass slide and allow it to dry at 20-25°C(room temperature).
- Place one drop of spore suspension on the same spot where fungicidal suspension was placed with the help of pipettes.
- Incubate the fungus by placing the slides Petri plates containing moist blotters for 24 hours at 20-25°C.
- Observe the germination of spores at six and 24 hours after setting of the experiment in low power/high power of the microscopes.
- Note the number of spores germinated in each concentration replication wise and calculate the percentage of spores inhibited from germination by using the following formula.

I= 100 (C-T)/C

Where:-

I = Inhibition percentage

C = No. of spores germinated in control

T = No. of spores germinated in treatment

Note:

i) The spore is said to be germinated if the length of the germ tube exceeds half the smaller diameter of the spore.

ii) Count at least one hundred spores for each concentration of test chemical.

Data obtained on the effect of different concentrations of the test chemicals and the germination of spores of test fungus is plotted on a graph paper to obtain 'Dose Response Curves.

Evaluation of fungicides and antibiotics against soil borne fungi *in vitro*:

Relative efficacy of fungicides or antibiotics is studied in vitro following poisoned food technique (Grover and Moore, 1961). The fungicides or antibiotics used in the experiment. Different concentrations i.e. 1, 10, 20, 50, 100, 200 and 500 ppm of these fungicides or antibiotics are taken on active ingredient basis. In order to prepare stock solution of fungicides or antibiotics, desired quantity of respective fungicides or antibiotic are added to measured quantity of sterilized distilled water and shaken gently. Known quantity of fungicides and antibiotics solutions are added to potato dextrose broth medium. The potato dextrose broth media without fungicides serve as control. Generally, three replications are kept for each concentration of fungicides tested. The Erlenmayer flasks (250 ml capacity) containing 100 ml media are inoculated with mycelial disc (5 mm diameter) of soil borne fungi taken from periphery of actively growing culture in Petri dishes and incubated at 25°C for seven days. The mycelial mat is filtered on previously weighed filter paper (Whatman No. 1) and dried for 48 hours at 60°C in the oven. The dry weight of mycelial mat in different fungicide treatments and control is recorded. The inhibition of mycelial growth is calculated.

Testing efficacy of fungicides or antibiotics in pot conditions:

The fungicides or antibiotics are used as seed treatment. The seeds are treated with the respective fungicides or antibiotics at 2 and 3 g kg^{-1} seed (proprietary formulation basis, w/w) before sowing. The desired amount of fungicides or antibiotics is added to 50 g seeds in 250 ml flask and shaken thoroughly for 4 to 5 minutes. Fungicides or antibiotics treated seeds are sown in earthen pots containing soil borne fungi inoculated soil. Five or ten seeds are sown in each pot. Suitable control is kept in which untreated seeds are sown in soil borne fungi inoculated soil. Three replications are kept for each treatment. The pots will be irrigated at suitable interval with uniform amount of water. Disease incidence will be recorded at 15 days interval up to 60 days after sowing.

$$\text{Disease Incidence (\%)} = \frac{\text{No. of infected plants}}{\text{Total no. of plants}} \times 100$$

10 IDENTIFICATION OF PLANT PARASITIC NEMATODE- HETERODERA

Materials: Stereoscopic binocular microscope, compound microscope, glass slide, cover slip, cover slip, glass wool, forceps, picking needle, alcohol, burner, nail polish, nematode suspension.

Procedure:

1 Prepare semi- permanent mount.

2 Fix the slide under compound microscope & identify it.

Systematic Position:-

Order	Tylenchida
Sub order	Tylenchina
Super family	Tylenchoidea
Family	Heteroderidae
Sub family	Heteroderinae
Genus	Heterodera, Globodera

Species of Heterodera -

i) avenae - Cereal cyst nematode(wheat & barley) found in north India

ii) zeae - Maize cyst nematode

iii) cajani - Pigeon pea cyst nematode (tur, mung & cowpea)

iv) oryzicola - Rice cyst nematode (rice & banana) found in Kerala, M.P., Orissa & West Bengal.

Morphological characters:-

i) **Body -** Slender in males (1.0-2.0 mm) and larvae (0.3-0.6 mm) In females, typically swollen, lemon shaped (0.5-0.8 mm)

ii) **Colour -** White or yellow, cyst dark brown, lemon shaped (0.8 mm long & 0.5 mm wide) or nearly same shape as that Meloidogyne female.

iii) **Stylet -** Short in males with rounded basal knobs & in larvae, more than 0.02 mm long.

iv) **Oesophagous -** With well-developed median bulb & lobe extending back & overlapping the intestine.

v) **Spicule -** Near the posterior end of males

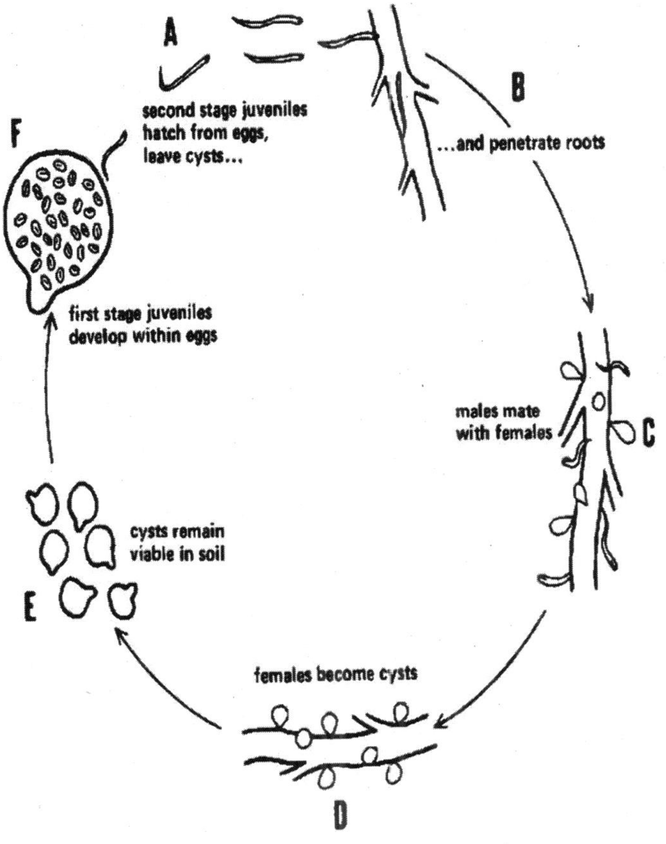

Cereal Cyst Nematode Disease Cycle

11 IDENTIFICATION OF PLANT PARASITIC NEMATODE-MELOIDOGYNE

Material:

Stereoscopic binocular microscope, compound microscope, glass slide, cover slip, teasing needle, forcep, watch glass, stained root of tomato.

Procedure:

1. Preparation temporary mount of nematode from stained root of tomato.

2. Fix the slide under compound microscope for identification of different stage/ stages of root knot nematode.

Systematic Position:-

Order	Tylenchida
Sub order	Tylenchina
Super family	Tylenchoidea
Family	Heteroderidae
Sub family	Meloidogyninae
Genus	*Meloidogyne*
Species	i) *incognita* ii) *javanica* iii) *arenaria* iv) *hapla*

Morphological characters:-

i) **Body** - Elongate larvae and male typically saccate, spheroid with a distinct neck in females.

ii) **Stylet** - In males, strong with rounded knob & in females, more slender than males.

iii) **Oesophagous** - With large median bulb followed by short isthumus.

iv) **Excretory pore** - Often seen with part of excretory tube in the area between posterior part of stylet knobs and opposite to median bulb.

v) **Vulvas & anus** - Females typically opposite to neck and surrounded by a pattern of fine lines like human fingerprint (Perennial pattern).

vi) **Spicule** - Very near the terminus of males bursa is absent.

Result: Following stage/ stages of *Meloidogyne* are seen

i) Egg stage	Oval shape
ii) 1st stage larvae	Within egg
iii) Second stage larvae	Inside root
vi) III stage larvae	Having spike
v) IV stage larvae	Spherical shape
vi) Adult female (with egg mass)	Swollen, round, gonad didelphic prodelphic large rectal glands secretes matrix.

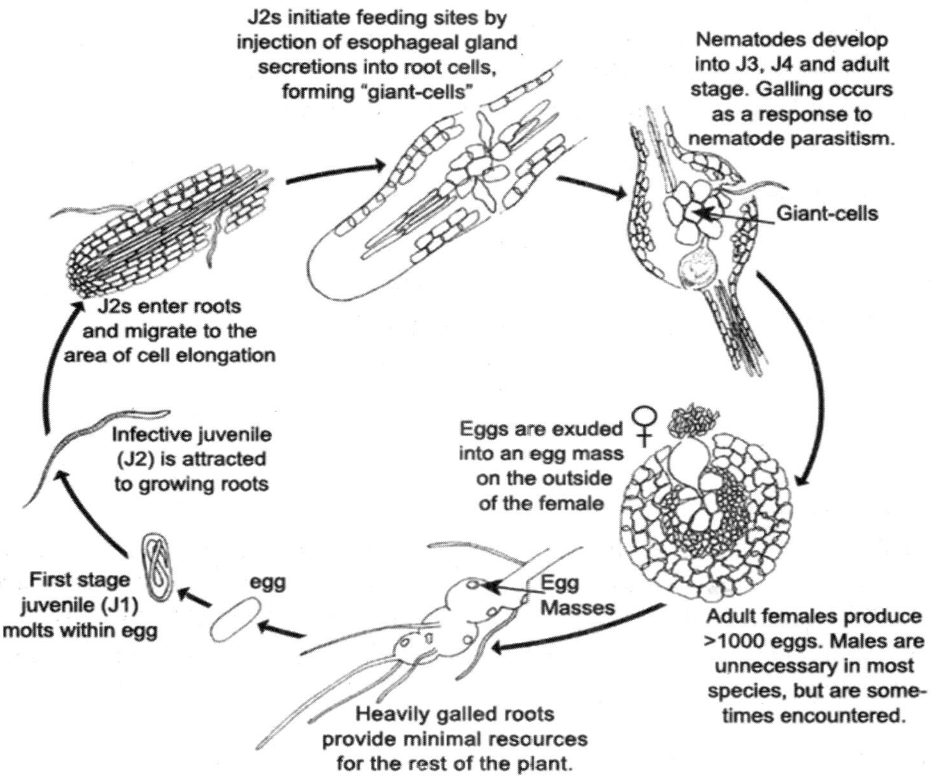

Root Knot Nematode Disease Cycle

12 IDENTIFICATION OF PLANT PARASITIC NEMATODE-ANGUINA

Materials: Stereoscopic binocular microscope, compound microscope, glass slide, cover slip, glass wool, forcep, picking needle, alcohol, burner, nail polish, nematode suspension.

Procedure:

1. Prepare semi- permanent mount.
2. Fix the slide under compound microscope & identify it.

Systematic Position:-

Order	Tylenchida
Sub order	Tylenchina
Super family	Tylenchoidea
Family	Tylenchidae
Genus	*Anguina*
Species	*tritici*

Morphological characters:-

Female:- Body obese, stylet short, medium oesophagealbulb with valvular apparatus, basal bulb enlarged and lobed. Ovary single , reflexed oocytes arranged in multiple rows around rachis. Vulva post equatorial. Tail elongated conoid.

Male:- Comparatively slender bodies testis single reflexed. Spicules well developed caudal alae cover nearly entire tail.

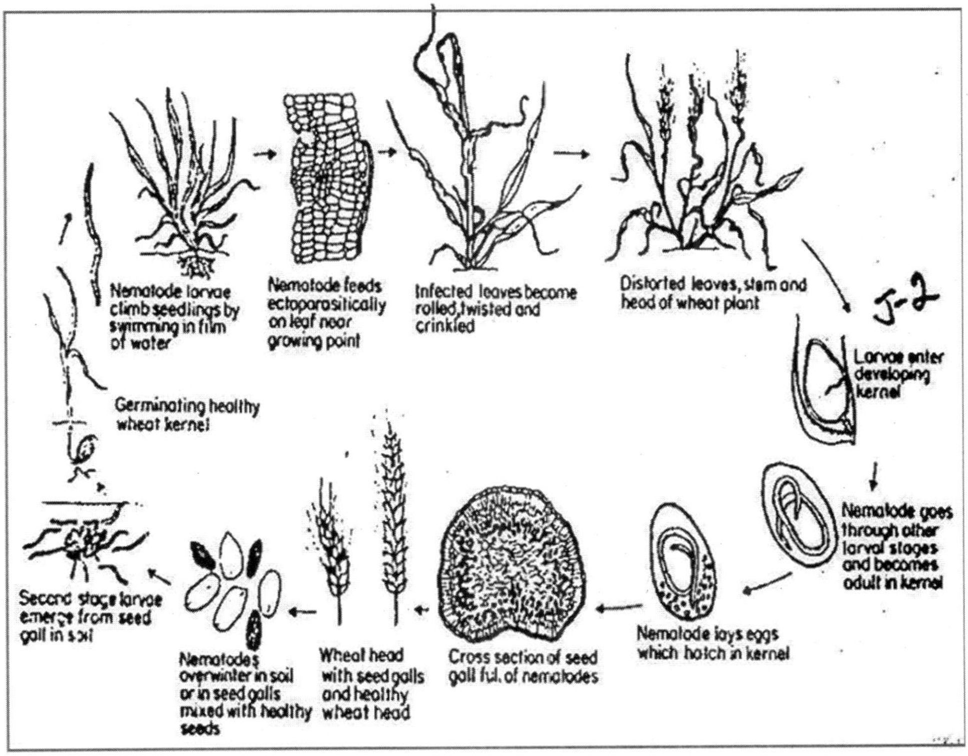

Seed Gall Nematode Disease Cycle

13 EXTRACTION OF NEMATODE FROM SOIL BY SIEVING AND DECANTATION METHOD

(BAERMANN'S FUNNEL ASSEMBLY)

Materials: Sieve set (16, 60, 100, 200 and 350 mesh), two plastic bowls, Baremann's funnel assembly, beakers (16 mesh sieve means 16 holes/inch2).

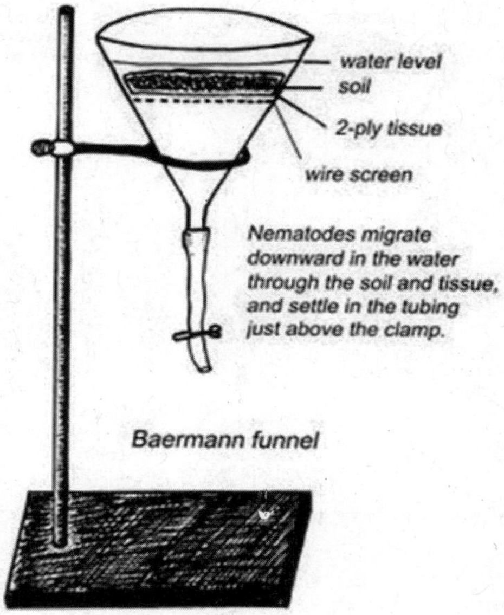

Baermann funnel

Procedure:

1. Mix soil sample thoroughly by shaking plastic bag.

2. A measured amount of soil (250 ml) volumetrically is taken in a bowl. The bowl is filled with water and a gentle shake is given.

3. The nematode being lighter will float on surface and soil particle settles down by gravity.

4. The above suspension is taken to another bowl through a sieve of 16 mesh and soil on mesh is discarded.

5. The process is repeated through a series of sieves of 60, 100, 200 & 350 meshes.

6. The material on above sieves are transferred in a beaker with the help of water current and finally on to a Baermann's funnel assembly and left for 24-48 hrs.

7. The nematode suspension is collected in a beaker through Hoffmann's clip.

Advantage:

The method allows recovery of most of nematodes from a large soil sample.

Disadvantage:

The method requires expensive sieves and trained worker.

14 SAMPLING AND EXTRACTION OF NEMATODES FROM DISEASED PLANT MATERIALS

Objectives: To collect migratory endoparasitic worm from roots

Materials: Stereo- microscope, Lactophenol, beaker, needle, gas stove, bowl, sieve (300 meshes), aluminium mesh, bloating sheet, Petri plate and nematodes infested roots.

Procedure:

A. From soft plant parts: Soak tender plant parts like stem, leaves, galls or thin roots in water for one or two hours. Tear the tissue apart using needles. Observe directly under a stereo- microscope. Aphelenchoides, Anguina, etc. are collected using this technique.

B. From harder plant parts:

- Blend these plant parts (eg. roots of perennial plants, cotton, etc.) in a mixer for 20 seconds. Wash the material on to a 300 mesh sieve with gentle stream of water. Collect the washing in to a beaker. Observe directly the contents under a stereoscope or proceed as follows.
- Transfer the contents as collected above on a moulded aluminium mesh lined with wet tissue paper. Put it over Petri-plate containing water. Observe the contents of the Petri plate after 24 hours.

15 PREPARATION OF NEMATODE MOUNTING

Materials: Lactophenol, acid powder, beaker, gas stove, bowl, bloating sheet.

Composition of Lactophenol

Lactic Acid	1 part	=200 ml
Phenol	1 part	=200 ml
Glycerol	2 part	=400 ml
Distilled Water	1 part	=200 ml

Dissolve 1 gm acid fuchsin powder in 10ml distilled water and add this solution to 1 litre lacto phenol. Fill the 0.1% acid fuchin in bottle.

Staining procedure:

- Wash the infested roots and remove the soil.
- Pour about 150 ml or 0.1% acid fuchsine solution to cover roots, into the beaker. Set the beaker into a hot plate and heat the staining solution almost to the boiling point.
- With the help of forceps, place a few roots at a time into the staining solution for about one minute or as long as it takes to stain roots.
- Rinse roots with cold water to remove excess stain and stop further heating.
- With forceps, place roots in a Petri dish.
- Cover roots with clear lactophenol solution.
- Observe under the microscope after 2-3 days to determine the progress of de-staining. Select root sections containing nematodes. Mount small piece in lactophenol on glass slide for observation under compound microscope.

Precaution:

1 Use exhaust fan during the staining of roots.

16 PLANT DISEASE DIAGNOSIS: THE COMPLETE PROCESS OF FIELD OBSERVATIONS, SAMPLE COLLECTION, SAMPLE PACKAGING AND LABORATORY DIAGNOSIS

Successful diagnosis of plant problems often depends on observation of symptoms, history of the problem and the site and knowledge of recent fertilizer, pesticide, or other chemical applications. Collect and package samples so as to provide representative, fresh samples of adequate quantity. If possible, collect whole plants with disease symptoms in early, middle and late stages of development.

Materials requirements:

- Plants with damage of some field crops or vegetables are recommended as samples, because whole plants can usually be collected. Also, plants showing leaf spot, blight, or crown or root lesions are suggested. Some common plant diseases that might be used for this exercise are anthracnose of chilli, beans; *Sclerotium rolfsii*—crown rot of pepper or other plants; Fusarium or Rhizoctonia—lower stem rot of bean; Alternaria leaf spot of marigold, cabbage, or tomato; and Botrytis blight of rose.
- Diagnostic questionnaire
- Plastic bags and twist ties
- Stereomicroscope and compound microscopes
- Microscope slides and cover slips
- Plastic bags and paper towels
- Two Petri dishes (10 cm dia.) of acidified potato dextrose agar

☞ Inspect a field, or landscape and locate damaged or diseased plants. For purposes of this exercise, fungal leaf spot diseases are preferred, but some lower stem (crown) or root rot diseases may also be successfully incorporated into the experiment.

☞ Thoroughly inspect the plant(s) for symptoms and signs. Make notes about the distribution of the disease—isolated, scattered, or the presence of a pattern.

☞ Obtain information on the site and problem history. Determine dates and rates of fertilizers, lime and pesticide applications and note recent weather conditions in the area. Complete a plant diagnostic questionnaire.

☞ Collect a sample of the problem. Collect whole plants if possible. If plants are small, collect several and if plants are 2-ft tall or more, one or two plants are sufficient. If it is not possible to collect a whole plant, collect several pieces of the plant showing varying stages of plant damage and symptom development.

17 PRESERVATION OF DISEASE SAMPLES

Preservation of diseased specimens is of much importance in the study of plant diseases. Such preserved specimens can be used as reference materials for comparisons. Dry and wet specimens of fungal and bacterial diseases can be preserved.

Collection and dry preservation of plant disease specimens

The main objective in preservation of plant specimens for herbarium use and storage is to dry the material quickly to prevent molding and minimize discolouration. The preserved specimen should remain in readily recognizable condition. The plant press is commonly used in the dry method. Plant or plant parts are arranged inside the folder newspaper, sheet with care that the materials remains flat and not variously folded. The newspaper is then placed between the blotters and the press is tightened. For drying it is placed in the sun or in a drying oven. When the plant is dry, it is placed in paper packets or mounted on herbarium sheets. The fleshy specimens can be dried in drier fitted with an electrical bulb.

Materials: Paper envelopes, plant press with blotters and newspapers

Procedure:

- ➢ Collect infected plant parts and place them in paper envelops. Use separate envelops for different kinds of diseased specimens and bring to the laboratory.
- ➢ Arrange (spread) the diseases specimens on the sheets of newspaper.
- ➢ Place the newspaper between the blotters and tighten the bolts of the press.
- ➢ Change the blotters at frequent intervals till the specimens become dry.
- ➢ Keep the press under sun for complete drying.
- ➢ Remove dried specimens from the press and keep them in non-absorbent tissue papers.
- ➢ Examine a few bits of specimens as early as possible after collection and identify the pathogen and label all the specimens giving the following information:
  a) Name of the student with I.D.No:
  b) Specimen number:
  c) Name of the host:
  d) Name of the causal organism (binomial):

Record:

Label all the specimens properly and prepare a list of specimens preserved Sketch of plant.

a. Wet preservation of diseased specimen

The fresh diseased specimen are treated with FAA solution and transferred to dilute (5%) formalin for storing the composition of FAA is given below:

Materials: Fresh diseased leaves, glycerol, vessel for heating, glass containers with screw tops.

Composition of F.A.A. (for 100 ml solution)

Ethyl alcohol (70%)	90 ml
Glacial acetic acid	5 ml
Formaldehyde/formalin	5 ml

Procedure: Formalin-acetic alcohol (FAA) is a good killing and fixing agent. After treatment with FAA, specimens can be stored in dilute (5% commercial) formalin. Virus affected green tissues, rust affected plant parts and leaf spots can be preserved in copper acetate solution.

b. Colour preservation of green plant parts, follow the following steps:

a) Thoroughly wash the specimen in running water
b) Place leaves in the boiling mixture containing 1 part of glacial acetic acid saturated with copper acetate crystals and 4 parts of water
c) Continue boiling until copper acetate replaces the natural green colour of cells, (approx. 15 min.)
d) Take out the material; wash thoroughly in running water again to remove excess copper.
e) Preserve the specimen in 5% commercial formalin in glass containers/ specimen jars with screw caps

For preservation of coloured fruits or stems of woody tissues, the prepare the following solution

- Zinc chloride : 50 g
- Formaldehyde (40%) : 25 ml
- Glycerol : 25 ml
- Water : 1 liter
- Wash and preserve the specimens as such in the above solution without boiling

MILESTONES IN THE DEVELOPMENT OF MOLECULAR PLANT PATHOLOGY

- 1988 K. Mullis and collaborates deviced the polymerase chain reaction (PCR). This technique for amplifying small amount of nucleic acid quickly has great impact on several fields of biology, including plant pathology.
- 1989 MB Dickman and PE Kolattukudy demonstrated that a cloned cutinase gene transforms a fungal wound pathogen in to one capable of directly penetrating the host cuticle. The experiment strongly supported the role of cutinases in direct penetration by certain fungi.
- 1989 the advent of Arabidopsis stimulated the search for mutants with altered disease resistance, plant functional genomics, and the Arabidopsis genomics sequencing project. This information will greatly affect disease control strategies in twenty first century.
- 1990 R Hain and coworkers demonstrated that tobacco plants transgenic for the phytoalexin synthesis gene, stilbene synthase, exhibit enhanced disease resistance. This work opened up a new strategy for obtaining disease resistance transgenic plants.
- 1990 the group of DF Klessig and IR Raskin and JP Metraux and J Ryals demonstrated that salicylic acid was associated with systemic acquired resistance the identity of the initial systemic signal, however it remain elusive
- 1990 B Kearney and BJ Staskawicz showed that the avrBs2 virulence gene of *Xanthomonas campestris* had an important virulence function. The finding was followed by other demonstration that avirulence genes might in fact be effectors for pathogen virulence and fitness that have been co-opted by plant disease resistance genes.
- 1991 PJGM DeWit and colleagues clone that first fungal avirulence gene, avr9, from *Cladosporium fulvum.*
- 1991 R Broglie and collaborator report that chitinase transgenic plants exhibit enhanced fungal disease resistance. This stimulates inserted in the idea of transgenic disease resistance plants.
- 1992 SP Briggs, JD Walton and colleagues clone the Hm-1 resistance gene from corn and demonstrate that its protein product detoxifiles the host selective toxin of *Cochliobolus carbonum*.
- 1993 GB Martin and collaborators clone Pto, the first plant disease resistance gene modeling the HR. Curiously, no other similar protein kinases have subsequently been shown to behave as resistance gene.
- FM Ausubel, BJ Staskawicz and collaborators molecularly clone Rps2, the first known leucine rich repeat plant disease resistance gene. Subsequently, dozens of similar genes are cloned several plants.

- 1995 Y Yang and DW Gabriel showed that Xanthomonas avirulence proteins of the avrBs3 family contain functional plant nuclear localization signals. This suggested that these proteins are introduced into plant cells, presumably by the bacterial hrp gene system, but are also translocated to and function in the nucleus.
- 1996 B Baker and colleagues demonstrated that the tobacco 'N' resistance gene against TMV function in transgenic tomato plants to provide viral resistance. This report encourages the idea that resistance genes may function in heterologus plant species.
- 1997 J Ryals and colleagues showed that the Arabidopsis NIMI gene (also called NPRI) has homology to the vertebrate-1-kb gene, known to be involved with innate immunity. The findings cement a linkage between active resistance in plants and animals and suggest that the two systems evolved from a common ancestor.
- 1999 F Reinach and associates finish complete genome sequencing of the first bacterial plant pathogen, *Xylella fastidiosa*. This work portrends greater interest in sequencing the genomes of a large number of species and strains of plant pathogens.
- 2000 r gene regulation and co-ordination of multiple loci understood in Arabidopsis.

18 MOLECULAR TECHNIQUES FOR DETECTION OF PLANT PATHOGENS

(A.) RAPD (Random Amplified Polymorphic DNA)

Principle of the PCR:

The purpose of a PCR (Polymerase Chain Reaction) is to make a huge number of copies of a gene. This is necessary to have enough starting template for sequencing

Cycling Reactions:

There are three major steps in a PCR which are repeated for 30 or 40 cycles. This is done on an automated cycler which can heat and cool the tubes with the reaction mixture in a very short time.

1.Denaturation at 94°C

During the denaturation, the double strand melts open to single stranded DNA, all enzymatic reactions stop (for example: the extension from a previous cycle).

2.Annealing at 54°C

The primers are jiggling around caused by the Brownian motion Ionic bonds are constantly formed and broken between the single stranded primer and the single stranded temple. The more stable bonds last a little bit longer (primers that fix exactly) and on that little piece of double stranded DNA (template and primer) the polymerase can attach and starts copying the template. Once there are a few bases built in the ionic bond is so strong between the template and the primer that it does not break any more.

3. Extension at 72°C

This is the ideal working temperature for the polymerase .The primers where there are a few bases built in already have a stronger ionic attraction to the template than the forces breaking these attractions. Primers that are on positions with no exact match get loose again (because of the higher temperature) and don't give an extension of the fragment.

The bases (complementary to the template)are coupled to the primer on the 3′ side (the polymerase adds dNTP's from 5′ to 3′ reading the template from 3′ to 5′ side bases are added complementary to the template)

PCR: polymerase chain reaction

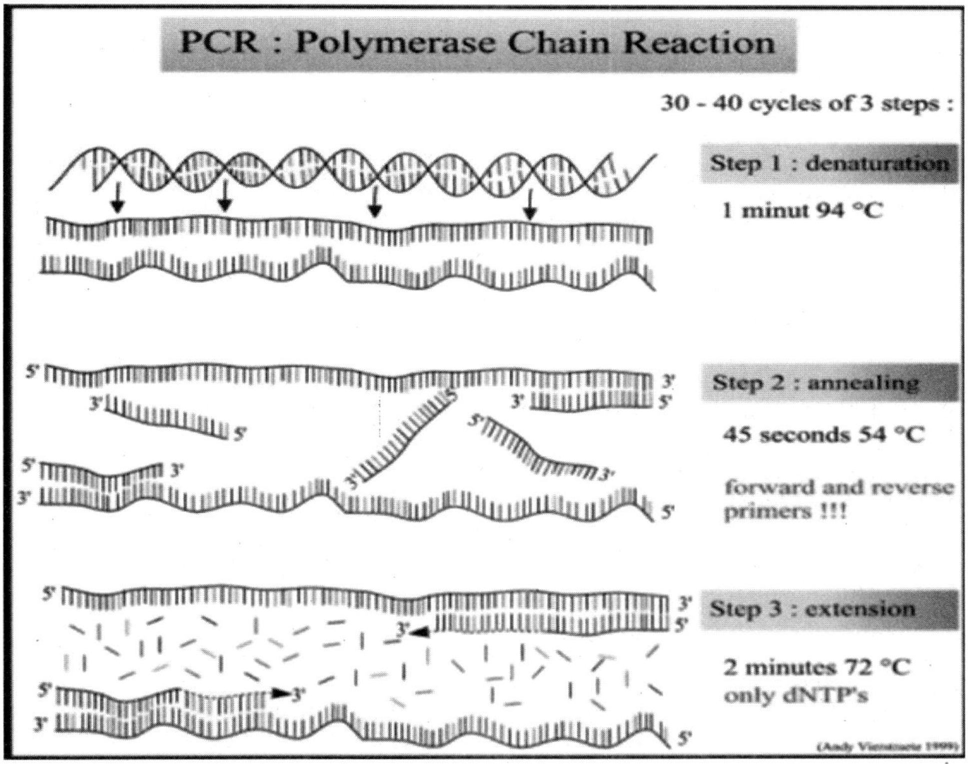

Steps to use PCR

- Load the reactions into 0.2 ml PCR (use eppendorf tubes to ensure good fit in machine)
- Make sure to use small holes in the PCR machine
- You can load up to 25 tubes at a time.
- Close lid and turn knob until it stops.
- Turn on PCR machine (switch on back)
- The menu should point at START (if not use arrows up and down)
- Press ENTER
- Use arrow keys to select the program you want to run
- Press ENTER
- The program will ask you what kind of tube you are using and the reaction volume; select the tube with the SELECT button enter the volume on the number keypad; hit ENTER

- ☞ While the program is running you can use the OPTION key to check how much longer you have to wait
- ☞ This is a great time to pour a gel

(B.) ENZYME LINKED IMMUNOSORBENT ASSAY (ELISA)

Introduction:

The Enzyme Linked Immunosorbent Assay (ELISA), first introduction into plant virology about 43 years ago (Vollar *et al.*, 1976), is one of the most widely used serological tests for plant viruses detection, identification, quantification and comparison owing to its simplicity, sensitivity (detection limits of 1-10 ng/ml) and adaptability. The amount of virus present in proportion to the amount of enzyme – labeled specific antibody forms the basis of the test.

Principle:

The double antibody sandwich ELISA (DAS-ELISA, direct procedure) has been the most popular ELISA procedure for plant virus detection. However, for those laboratories engaged in the detection of large number of viruses, another form of ELISA, i.e. direct antigen coated ELISA (DAC-ELISA, indirect procedure) is preferable, wherein the immune probe (specific/primary antibody) is not enzyme-labeled but itself identified by another enzyme-labeled probe (anti immunoglobulin molecules, secondary antibody).

Protocol:

- Add 200 µl test/control samples extracted/ diluted in coating buffer containing 2% (w/v) PVP to each well of the microtitre plate. Cover the plate and incubate at 37°C for ½-1h.
- Empty and wash the plate by flooding the wells with PBS-T for about 3 min. repeat wash and soak operation thrice, then empty plate and shake out residual liquid draining on a paper towel.
- Add 200 µl blocking solution (heterologous protein) to each well. Incubate at 37°C for 1h to back polystyrene well reactive surface.
- Wash the plate three times as in step above.
- Add 200 µl crude antiserum (primary antibody) diluted PBS-TPO to each well. Incubate at 37°C for 2h or overnight at 4°C.
- Wash the plate three times as above.
- Add 200 µl freshly enzyme-labeled (Alkaline phosphatase ALP) antirabbit lgG (secondary antibody) diluted in PBS-TPO to each well. Incubate at 37°C for 1h or overnight at 4°C.
- Wash the plate three times as above.
- Add 200 µl freshly prepared substrate (paranitro phenyl phosphate PNPP 0.5-1 mg/ml) to each well. Incubate at room temperature 1-2h or as long as necessary to observe reaction.

- Terminate reaction (if necessary) with 50 µl 3M NaOH to each well.
- Access enzyme mediated colour reaction either by visual observation or by measuring absorbance at 405 nm. Samples are recorded as infective or positive whose ELISA valves exceed the healthy sample valve by 2 or 3 times.
- Draw test plan in checker board format and determine optimum dilutions of various reagents to be used.
- Include both positive and negative reference samples to validate the assay

Anticipated Result:

After addition of substrate yellow colour develops which can be scored either visually or in ELISA reader. Background colour obtained with healthy sample will give some absorption at 405 nm. A_{405} valve twice that of healthy sample is taken as indicative of presence of the pathogen. A minimum of three wells for each dilution, two dilutions per sample are required to draw conclusion.

(C.) DOT- IMMUNOBINDING ASSAY (DIBA)

Introduction:

It is simple dot-blot immunoassay wherein antigen suspensions (1-5 µl each) are applied directly to the nitrocellulose/nylon membranes without prior electrophoretic separation. It is also referred dot-ELISA i.e. ELISA on membranes. It is less time consuming, expensive and labour-intensive than conventional ELISA procedures and has been used to detect a variety of plant viruses.

Principle:

In this procedure the test is carried out on a membrane. It differs from ELISA in having insoluble substrate reaction that gives rise to colour development.

Protocol:

- Extract tissue 1:10 (w:v) in antigen extraction buffer and expressed through cheese cloth.
- Pipette 0.8 ml expressed sap into a 1.5 ml eppendrof tube. Add 0.4 ml $CHCl_3$ vortex and centrifuge (12000 g) for 2 min.
- Pipette 200 µl clarified sap (upper aqueous layer) in to 800 µl antigen extraction buffer and vortex.
- Put on the gloves, cut a desire size piece of NCM and draw a lattice of squares of 1X1 cm each with a soft-lead pencil. Always use forceps for handling the membrane.
- Wet the NCM by floating it on TBS and then air dry. Spot the test/control samples (5-10 µl) by hand.
- Allow NCM to air dry and immerse in blocking solution with gentle oscillation for 1h at room temperature or overnight at 4°C.
- Rinse once in TBS for c. 10 min.

- Incubate for 1h at room temperature or overnight at 4°C in crude antiserum (primary antibody) diluted in TBS-SDM.
- Rinse thrice in TBS for c. 10 min. each
- Incubate for 1h at room temperature or overnight at 4°C in secondary antibody (enzyme linked ant rabbit lgG) diluted in TBS-SDM.
- Rinse thrice in TBS for c. 10 min. each
- Incubate in substrate solution at room temperature in the dark and watch for colour development (good purple colour development take 5-10 min.)
- Rinse thrice in the fixing solution for c. 10 min. each and then air dry NCM between whatman filter paper sheets.
- Assess the result by visual observation by comparing the intensity of purple colour or by using densitometer.
- Photograph NCM (dry or wet) and store protected from light.

Anticipated Result:

A light pink to dark violet colour develops in the samples having the pathogen. No such colour development is seen with healthy samples. It is more qualitative test to detect the pathogen. Crude extract of a large number of samples can be dotted, stored and used.

(D.) IMMUNOSORBENT ELECTRON MICROSCOPY (ISEM)

Introduction:

Immunosorbent Electron Microscopy (ISEM), developed by Derrick (1973), is a valuable technique which combines the specificity of serological test with the possibility to visualize the type of viral antigen in the electron microscope. The technique not only helps in the detection of virus particles but also facilitates virus identification.

Principle:

The virus specific antibody is initially coated on the grids on which samples are mounted. The viruses present in the sample get trapped and concentrated in one place making it easy visualize. If it is further treated with antiserum of related virus, the binding of antibody gives rise electron dense "decoration" around the particles.

Protocol (trapping and Decoration):

- Pipette a drop of crude antiserum (c. 20 μl) on parafilm membrane in a moist Petri dish and float a grid with coated side on the drop. Incubate the Petri dish for approx. 30 min. to 1h at room temperature
- Wash the antiserum coated grid (ACG) in phosphate buffer for 10 min.

- Pipette a drop (c. 20 µl) of extract from diseased tissue and float the grid on it. Incubate for 30 min to 2h.
- Wash the grid with buffer for 10 min.
- Pipette a drop (c. 20 µl) of 1:50 dilution of antiserum and placed the grid on it. Incubate at 37°C for 30 min to 1h.
- Wash the grid with water followed by 2% aqueous urinal acetate. Drain and dry the grid before observing under EM.
- Observe the number of virus particles trapped as compared to the leaf dip and their decoration by the antibodies

Anticipated Result:

- In simple antibody, since virus particles get trapped, even if it is present in very low concentration, it can be visualized easily. If the second antibody used is for any related virus strain, uniform electron dense decoration can be seen around the particle. If there is mixture of two viruses of similar morphology, they can be easily distinguished by decoration.

(E.) ISOLATION OF FUNGAL GENOMIC DNA

Principle:

Good quality fungal DNA can be prepared by a variety of methods. If the DNA is to be used for the construction of genomic libraries or in southern blot analysis, it is essential that the DNA is intact and of a high molecular weight (between 100-200 kb). The size and integrity of the DNA can be assessed by slow, low voltage agarose gel electrophoresis in TBE or TAE buffer.

Requirements:

Chemicals required: Chloroform, Isoamyl alcohol, Isopropanol, NaCl, EDTA, Tris HCl, PVP, CTAB, β mercaptoethanol, Ethanol, Liquid Nitrogen

Glassware's & Accessories: Reagent bottles, Centrifuge tubes, Beakers, Micropipettes, Spatula, Mortar & pestle

Protocol:

Harvesting fungal tissue- Exponentially growing fungal cultures in liquid medium (Potato dextrose broth) by shaking for 5-7 days at 25°C, yield the greatest amounts of good quality DNA. Mycelium is collected by filtration through sterilized filter paper lining a Buchner funnel and then extensively water washed. The mycelium is blotted dry between the layers of tissue and immediately frozen in liquid nitrogen in full packets. The tissue is gently broken into fine pieces by crushing the foil envelopes with a pestle. Care must be taken not to allow the tissue to thaw as this will cause lysis and the release of endogenous nucleases. Frozen tissue can be stored at -70°C for long period of time (>1 year).

DNA extraction:

CTAB method is routinely used for the preparation of fungal DNA which has been used to prepare high molecular weight DNA from a range of Basidiomycetes, Ascomycetes, Deuteromycetes and Oomycetes fungi.

Steps:

- Pre-cool a pestle and mortar at 4°C. Grind 5g mycelium to a fine powder in liquid nitrogen and transfer to a plastic sterile tube, ensuring that the tissue does not thaw.
- Add to it 10 ml of pre warmed isolation buffer (CTAB buffer) and incubate for 60 min with occasional stirring.
- Extract for 10 min with equal volume of chloroform: isoamyl alcohol (24:1).
- Centrifuge at 11,000 rpm for 20 min at room temperature (24°C).
- Seperate the aqueous phase and transfer to a fresh tube.
- To aqueous phase add 0.6 volume of ice cold isopropanol and 0.1 volume of sodium acetate, and incubate at -20°C for 1hr or overnight at 4°C.
- Centrifuge at 10,000 rpm for 10 min at 4°C, discard the aqueous phase.
- Wash the DNA pellet obtained with 70% ethanol (5 ml).
- Centrifuge at 8,000 rpm for 10 min at 4°C, discard the aqueous phase.
- Dry the DNA pellet and dissolve in 200 μl of TE buffer or sterile distilled water.
- Genomic DNA dissolves slowly in TE following ethanol precipitation, particularly if the concentration exceeds 1 μg/μl. It may be necessary to resuspend the DNA on a rotary table at 20 rpm overnight at room temperature. After washing with 70% ethanol, the DNA pellet, which is obtained, should look like white, thick thread like mass. This DNA obtained is further quantified spectrophotometrically and agarose gel electrophoresis.

Composition of CTAB (Cetyl Tri methyl Ammonium Bromide) buffer: (500 ml)

CTAB (2%): 10 g

NaCl (1.4 M): 40.95 g

Tris-HCl (1 M; pH-8.0): 50 ml

EDTA (0.5 M, pH-8.0): 20 ml

PVP

Makeup the volume to 500 ml and add 0.2% β-merceptoethanol before use.

Purification of DNA (RNase Treatment) (Optional):

- Add 5 μl RNase in 100 μl of DNA
- Incubate at 37°C for 30 minutes in water bath.

- Add equal volume of Chloroform: Isoamyl alcohol (24:1) to a vial.
- Centrifuge at 10,000 rpm for 10 minutes at room temperature.
- Take out the supernatant and add double volume of chilled isopropanol + 1/10 vol. of sodium acetate (3 N) to the supernatant.
- Incubate the vial at -20°C for 1h.
- Centrifuge at 12,000 rpm for 15 minutes at 4°C.
- Wash the pellet with 70% ethanol by centrifuging at 6000 rpm for 5 minutes.

Role of chemicals:

1. CTAB – Cetyl Trimethyl Ammonium Bromide

   ☞Disrupts cell structure.

   ☞Causes precipitation of lipids, proteins and polysaccharides.

   ☞Complexes with DNA (keeps DNA in solution at high concentration).

2. 1.4 M NaCl – Keeps Ionic concentration of buffer high (DNA in the solution)

3. 20 mM EDTA – Ethylene Diamine Tetra Acetate

Chelator (Cofactors), Inactivates Nucleases (RNAases & DNAases)

4. Tris HCl, pH 8.0 – Optimises pH for for DNA & RNAExtraction.

5. 0.2% β- mercaptoethanol – Antioxidant, Prevents oxidation of Polyphenols

6. $CHCl_3$ – Deproteinization, forms organic phase, removes lipids, polysaccharides.

8. Isopropanol– precipitates DNA

9. 70% Ethanol

   ☞ Washing DNA pellet

   ☞ Removes Salt

(F.) RESTRICTION FRAGMENT LENGTH POLYMORPHISM (RFLP)

Introduction:

RFLPS are difference in the length of specific restriction fragments generated with a particular restriction enzyme. It is called as DNA finger printing. In RFLP, the polymorphism can occur by possessing a new cutting site for restriction enzyme cleavage or a cutting site is lost. Different steps involved are;

(a) Restriction digestion of DNA

(b) Gel electrophoresis of restricted fragments

(c) Southern blotting

(d) Hybridization with a specific probe and

(e) Detection by Autoradiography.

Major applications:

1. To produce a fingerprint-use for identification.
2. Linkage analysis where RFLPs can as genetic marker associated with resistance genes.
3. Combination of restriction enzyme and probe, radioactive bands from Southern blots will be unique species.

A. Restriction of genomic DNA by Restriction endonucleases

Principle:

The restriction/modification system in bacteria is a small scale immune system for protection from the infection by foreign DNA. Enzyme cleaves the DNA at a palindromic sequences called as Restriction endonuclease [they restrict foreign DNA within the cell].

- Restriction endonuclease are-
  - One of the primary tool in the modern molecular biology for DNA manipulation
  - Commonly named after bacteria from which they are isolated.
- The endonucleases hydrolysis is a spontaneous reaction and does not require addition of ATP.
- The reaction buffer usually has a buffer component (10mM Tris buffer, pH 8) 10 mM $Mgcl_2$, 1mM diethiothrietol, DDT 100μg/ml bovine serum albumin, BSA) and salt (Nacl).
- The biggest determinant of enzyme action is the ionic concentration [Nacl] of the buffer.

Protocol:

- Set up a 100 μl restriction digestion of the fungal genomic DNA sample by adding the following to a 1.5 ml eppendrof tube.

Fungal genomic DNA	20 μl (10 μg)
SDW	24
Restriction buffer	5 μl
Restriction	1 U (5-10 U)
Total	50 μl

- Give a pulse spin, so that the contents are mixed well and brought to the bottom of the tube.
- Incubate at 37°C for 2h.
- After 2h take 5 μl of the restricted sample.

- Add 4 μl of loading dye, 10 μl of water.
- Mix by tapping, then give pulse spin to bring all the ingredients to the bottom of the tube. Now the sample is ready to be loaded in the gel.
- Check it on gel electrophoresis. Uniform fluorescence along the lane is indicative of good restriction.

From the trial run, based on fluorescence, a rough estimate of concentration of DNA in each sample will be obtained. Accordingly, aliquots of restricted samples can be taken for final gel running.

B. Gel electrophoresis of restricted fragment

- Prepare 1 X TAE solution from 50 X TAE stock solution
- Weight enough arose to make 0.8 solution in 1 X TAE buffer
- Melt agarose by boiling.
- Cooled at 40°C add Et Br to a final concentration of Etbr.
- Pour the agarose onto the gel casting tray which has been set previous.
- After agarose has set pour buffer so that the gel gets submerged.
- Load the sample into the wells along with a molecular weight marker.
- Connect electrical leads to power supply.
- Run the electrophoresis for overnight.
- Turn of power supply.
- Put the gel on a UV transilluminatory and with a safety shield, visualize the gel.

C. Southern transfer

Principle:

- Transferring the DNA from the gel to a Nylon/Nitrocellulose was devised by Southern.
- The DNA would be transferred from the gel after being denatured and neutralized, through the buffer by capillary action onto the Nylon membrane. The DNA on the membrane should be immobilized and can be used for further analysis like hybridization.
- Once transfer has taken place to the membrane, it becomes a replica of the gel. The DNA should be denatured to make it single stranded, to allow the probe to bind to it.

Protocol:

- Trim off the agarose gel above the loading wells, the regions not having any DNA to fix the orientation , make a slanting cut on the left hand corner.
- Take measurement of the gel.

- Soak the gel for 10 min in volumes of 0.2N HCl. This step is called depuration. By this DNA is partially hydrolyzed, can pass through the gel even if it gets dried. The resulting fragments get transferred efficiently. Avoid depuration for a longer time.
- Rinse it in deionizer water.
- Soak the gel in a tray with denaturation solution with constant gentle agitation.
- After 45 minute, pour the denaturation solution out and add neutralization solution.
- While the gel is soaking, cut the membrane as per the gel measurement, trim off the membrane on the upper left hand corner.
- Soak the membrane in SDW for 2 min.
- Soak the membrane in 10 X SSC for 15-20 min.
- Pour transfer buffer in a tray and set the transfer.
- Next day, remove the assembly with the forceps.
- Soak the membrane in 6 X SSC for 5 min. to remove the agarose pieces.
- Dry the membrane. Immobilize the DNA by baking, at 80°C for 2h.

Hybridization with a probe

D. Probe preparation

Principle:

DNA is labeled by Random primer labeling.DNA fragment of fungal genome, cloned in earlier session will be used to make the probe. Probe will be made by random primer labeling method. In the presence of klenow, new strands are synthesized, which will have incorporation of radioactive dNTP.

Protocol:

Denature the DNA at 100°C for 5 min. immediately mix the following in (μl)

Denatured DNA	1
Primer [Random hexanucleotides] (100 ng/ μl)	1
10 X labelling buffer	5
dNTP [10 mM dATP, dGTP, dTTP]	2
Ditiothreitol [DTT] [20mM]	2.5
Nuclease free BSA	2
dCTP [4000 ci/mmol]	1
Klenow [DNA Pol l] [3U]	1
SDW	34.5
Total	50 μl

Incubate at 37°C for 2 h

E. Pre hybridization:

- The spaces in the membrane where DNA is not bound should be blocked to avoid non–specific binding of the probe. This is done by immersing the membrane in a solution containing large molecular weight DNA or RNA along with detergent and dispersing agent.
- Keep the membrane in one of the large glass cylinders of the hybridization oven. The yeast mRNA acts as a blocking agent. Cylinders are rotated to facilitate uniform spreading of a solution.

F. Hybridization:

- After 4h add the probe into the prehybridisation solution.
- Before adding the probe, it should be denature by boiling.
- Normally, probe is added at 0.5×10^6 cpm/ml of hybridization solution.
- Overnight incubation at 65°C.

Useful facts:

- A combination of temperature and salt concentration should be chosen for studying the hybridization.
- Nucleic acid hybridization is used in the detection and diagnosis of plant pathogens.
- The washing condition is considered as stringent when the temperature of washing is >65°C and salt concentration is very low (0.2 SSC).
- Nucleic acid hybridization is also used in the detection and diagnosis of plant pathogens.

Washing the blots:

- The next day morning pour hybridization solution out and add 30 ml washing solution. This is done to remove any unbound probe.
- Rotate the cylinders for 15 minutes at 65°C.
- Repeat the washing two more time.
- Take the membrane out by forceps and dry it out on a paper towels.

Detection by Autoradiography:

The particle emitted from a radioactive atom activates a photographic emulsion, much like light or X–ray. If the photographic emulsion is brought into close contact with a radioactive source, the particles emitted by source leave tiny, black silver grains in the emulsion after photographic development.

G. AMPLIFIED FRAGMENT LENGTH POLYMORPHISM

Introduction:

DNA fingerprinting refers to a set of DNA fragments on an electrophoretic gel. The banding pattern of the DNA fragments obtained either by restriction or PCR amplification with a specific or random primer will represents a particular isolate. There are many techniques available for the DNA finger printing viz., RFLP, RAPD, AP-PCR, etc. The choice of selecting a finger printing technique will depend on the purpose like gene mapping, population studies etc. Many DNA finger printing techniques have been developed which are generally based on one of the two strategies i.e. classical hybridization based finger printing viz., and RFLP and PCR based finger printing like RAPD and AP-PCR, AFLP technology incorporates both the techniques.

Principles:

The AFLP technique based on the amplifications of subset of genomic restriction fragments using PCR. DNA is cut with restriction enzymes, and double stranded adapters are ligated to the end of the DNA fragments to generate templates DNA for amplifications. The sequence of the adapters and the adjacent restriction site serve as primer binding sites for subsequent amplification of the restriction fragments. During PCR only those restriction fragments in which the nucleotides, will be amplified. The subset of amplified fragments is then analyzed by denaturing PAGE to generate fingerprints.

The restriction fragments for amplification are generated by two restriction enzymes, a rare cutter and a frequent cutter. The AFLP procedure results in predominant amplification of those restriction fragments which have a rare cutter sequence on one end and a frequent cutter sequence on the other end. The number of fragments to be amplified is reduced by using rare cutter, since only the rare cutter/frequent cutter fragments are amplified. This limits the number of selective nucleotides needed for selective amplification. AFLP is a tool that allows differentiation between individuals, genotypes and strains and the assessment of genetic diversity and phylogeny. Originally developed for genetic mapping in plants, this technique has characteristics of an ideal system for detecting genetic variation. The AFLP markers are `natural and are generated from a large number of independent loci from different parts of the genome the profiles generated are reported to be highly reproduction because of the high stringency PCR, due to the nature of the primers AFLP analysis involves selective amplification of fragments from restriction enzyme digest of genomic DNA. The genomic DNA is digested simultaneously with a hexa nucleotide cutter EcoR1 and tetra nucleotide cutter mse /, these results in more than 1.5 lakh fragments from a fungal genome of approximately 5×10 bp. Due to primer design and amplification strategy, these EcoR-1Mse1 fragments are prefer entiallyamplified. Following heat inactivation of the restrictions endonucleases, the genomic DNA fragments are ligated to EcoR-1 and Mse1 adapters to generate templates DNA for amplification. These common adapter sequences flanking variable genomic DNA sequences serve as primer binding sites on these restrictions fragments. Using this strategy, it is possible to

amplify many DNA fragments without having prior sequences knowledge.
PCR is performed in to consecutive reactions, in the first reaction, called pre amplification, genomic DNAs are amplified with AFLP primers each having one selected nucleotide. The PCR products of the pre amplification reaction are diluted and used as a template for the selective amplification using two AFLP primers, each containing three selective nucleotides selective primer is 33P-labeled before amplification. This two-step amplification strategy results in consistently cleaner and more reproducible fingerprints with the added benefit of generating enough templates DNA for thousands of AFLP reactions.
The most important factor in determining the number of restriction fragments amplified in a single AFLP reaction is the number of selective nucleotides in the selective primers. The selective primers in the AFLP analysis system I contain three selective nucleotides. In practice, using the AFLP analysis system I with plants having genome ranging in size from 510 to 610 bp, the number of amplified fragments per sample per prime pair averages 50, but may range from as 10 to100 depending on the sequence context of the selective nucleotides, and the complexity of the genome.
A second factor in determining the number of amplified fragments is the C and G composition of the selective nucleotides. In general, the fewer DNA fragments amplified. Also, the smaller the genome being analyzed, the fewer fragments amplified and the simple the fingerprint. The AFLP method is more robust and reliable than RAPDs, because more stringent reaction conditions are used, but it is more difficult and expensive to do.

Protocol:

A. DNA isolation :

For AFLP, high quality DNA is required. For this experiment DNA prepared in early experiments will be used. The kit AFLP I from GIBCIO-Bel will be used for the present study.

B. Restriction digestion of genomic DNA :

Following components are added to 1.5 ml micro centrifuge tube.

5X reaction buffer	- 5 µl
Sample DNA	- 18 µl
EcoR1/Msel	- 2 µl

Total volume made up to 25 µl by addition of sterile distilled water.

- Mix gently and then spin down in centrifuge.
- Incubate this mixture for 2 hr at 37°C.
- Keep the reaction mix ice and spin down and then place the tube in a water bath at 70 centigrade for 15 minute heat inactivate the enzymes.

C. Ligation of adapters:

- To the restricared DNA from previous step add the following components.

  -Adapter ligation solution -24 µl

  -T_4 DNA Ligase- 1 µl
- Mix gently, spin down and incubate at 20± 2 for 2h.
- The ligation mixture has to be diluted by adding 90 µl of TE buffer to 10 µl of reaction mixture.

D. Pre amplification:

- In 0.5 ml thin walled micro centrifuge tube following components are added.

Diluted template DNA	5 µl
Primer E-O	2.7 µl
Primer M-O-	12 µl
10 X PCR buffer	5 µl
Taq DNA polymerase	1 µl

Distilled water is added to make the final volume to 51 µl

- Mix gently and spin down and proceed for PCR
- The product is diluted 1-1:50 by adding 147 µl TE buffer to 3 µl PCR products.
- Unused portion can be stored at -20°C.

E. Primer labeling:

Primer labeling is done by phosphorylating the 5 end of EcoRI primers with [P^{32} or P^{33} r-ATP and T_4 Kinase]

The following components are added to a 1.5 micro centrifuge tube

EcoRI primers	18 µl
5X kinase buffer	10 µl
(r^{33} P) ATP (300°Ci/mmol)	20 µl
T_4 polybnucleotide kinase	2 µl
Total volume	1 µl

- Mix gently and spin down.
- Incubate the mixture at 37°C for 1 h.
- Heat inactivates the enzyme at 70°C for 10 min. and spin down in centrifuge.

F. Selective AFLP amplification:

1. For each primer pair, the following components are added in a 1.5 micro centrifuge tube mix l.

Mix I

Labeled ecori primer	5 μl
MSC 1 primer	45 μl
Total	50 μl

Mix ll

Distilled water	79 μl
10 X pcr buffer	20 μl
Taq DNA polymerase	1 μl
Total	100 μl

2. For the AFLP reactions following components are added to the thin walled microcentrifuge tubes

Per amplified DNA templates	5 μl
Mix l	5 μl
Mix ll	10 μl
Total	20 μl

(c) Mix gently and spin down

G. PCR amplification cycles:

- One cycle at 94°C for 30 s

  65°C for 30 sec.

  65°C for 30 sec.

  72°C for 60 sec.

- Lower the annealing temperature each cycle at 0.7 c during 12 cycles. This gives a touchdown phase of 13 cycles. Perform 23 cycles at

  94°C for 30 sec.

  56°C for 30 sec.

  72°C c for 60 sec.

  Total time 2 hrs 2 mins.

H. Gel analysis:

(a) Preparation of gel:

(i) Cleaning of plates

- Plate 1
- Plate 2

(ii) The spacer are put between two plates at the gride edges and the plates are sealed from sides and bottom by adhesive tape.

(iii) 5% or 6% polyacryamide gel is prepared for 100 µl

50 g urea

5.7 g acryamide

0.3 g bis acrylmide dissolved in 10 µl of tbe buffer and 40 ml water at 65°C

Total volume is made by distilled water for 100 µl and filtered. APS (500 µl)and timed (1.0 µl) are added just before casting the gel.

(iv) After casting, the adhesive tape is removed and the gel plates are fitted in the apparatus.

(b) Gel running:

- After PCR added an equal volume of dye to each solution.
- Heat the samples for 3 min. at 90°C and place immediately after on ice.
- Pre electophase the gel at constant power for 10 min.
- Stop the run wash the wells.
- Load 3 µl of each sample in the gel.
- Electrophase at constant power until the dye front touches down the bottom
- Separates the plates after removing it from gel apparatus.

Anticipated Results:

The banding pattern of the DNA will depend on the primer set used. The observation has to be made for the polymorphism of the binding pattern of the related isolates. The scoring is done for the presence or absence of a band. For a band present we score 1 and for absence of corresponding band in other lane will score 0. This data when fed to suitable software gives the relativity of the isolates.

CENTRIFUGE:

The centrifuge works using the sedimentation principle, where the centripetal acceleration causes denser substances to separate out along the radial direction (the bottom of the tube). By the same token lighter objects will tend to move to the top (of the tube; in the rotating picture, move to the center).

Process or Protocols for centrifugation

Typically specify the amount of acceleration to be applied to the sample, rather than specifying a rotational speed such as revolutions per minute. This distinction is important because two rotors with different diameters running at the same rotational speed will subject samples to different accelerations. During circular motion the acceleration is the product of the radius and the square of the angular velocity *r*, and the acceleration relative to "*g*" is traditionally named "relative centrifugal force" (RCF). The acceleration is measured in multiples of "*g*" (or × "*g*"), the standard acceleration due to gravity at the Earth's surface, a dimensionless quantity given by the expression:

A 19th century hand cranked laboratory centrifuge.

$$\textbf{RCF} = \frac{rw^2}{g}$$

Where

g is earth's gravitational acceleration,

r is the rotational radius,

w is the angular velocity in radians per unit time

This relationship may be written as

$$\text{RCF} = 1.11824396 \times 10^{-5}\, r_{\text{cm}} N^2_{RPM}$$

Where

r_{cm} is the rotational radius measured in centimeters (cm), and

N_{RPM} is rotational measured in revolutions per minute (RPM).

Types by intended use

- Ultracentrifuges are optimized for spinning a rotor at very high speeds and are popular in the fields of molecular biology, biochemistry and polymer science. This type may include preparative or analytical, fixed-angle or swing head varieties.
- Hematocrit centrifuges are used to measure the percentage of red blood cells in whole blood.
- Gas centrifuges, including Zippe-type centrifuges

SPECTROPHOTOMETER:

The spectrophotometer is an analytical machine that measures the transmission of a beam of light through a liquid sample (% Transmission) or the amount of light that does not pass through a liquid (extinction). The spectrophotometer uses specially designed sample containers (called cuvettes) to hold the liquid sample in the beam of light.

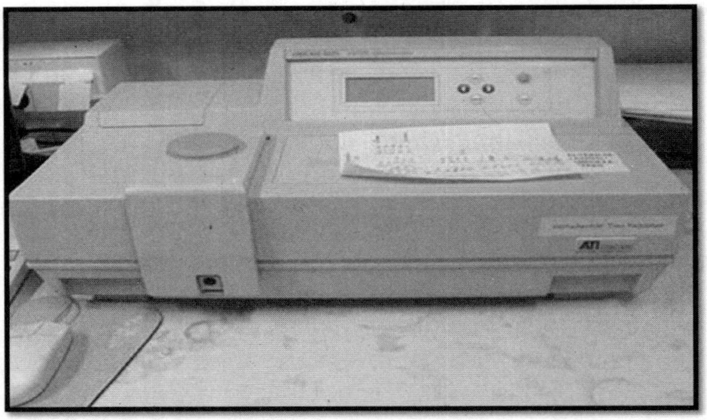

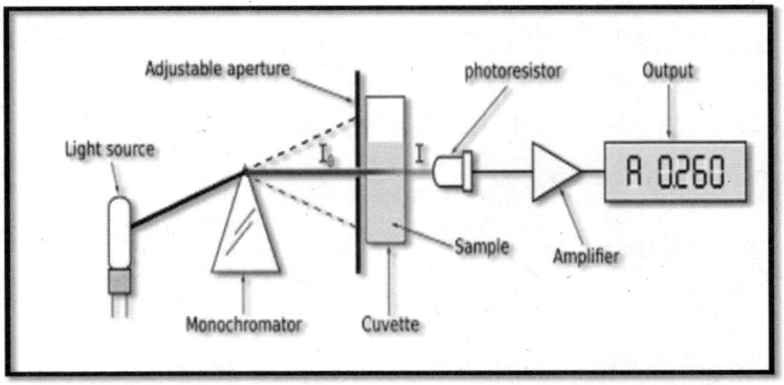

The wavelength of the beam of light may be varied for different applications, which may include measuring the:

- Turbidity ('murkiness') of a liquid
- Appearance or disappearance of a coloured compound during a chemical reaction
- Intensity of colour that may be related to concentration of a solute
- Wavelength of the colour in a liquid.

The spectrophotometer that we will use has the following features:

- A single sample (cuvette) holder
- Is used for only two applications

    Measuring the turbidity of a bacterial suspension to determine the concentration of bacterial cells (used to check the calibration of the spectrophotometer)

    Measuring apple pulp browning enzyme levels

- Is set to a wavelength of 686 nm or 590 nm (nm = nanometre = 10^{-9} metre) respectively for each of the applications
- Reports in % Transmission (%T)
- Is of a simple manual design and is not automated.

The spectrophotometer is set-up according to an in-house SOP, calibrated by zeroing against air (i.e. no sample cuvette in the cuvette holder) and setting 100%T against a 'blank' solution.

Basic Principles:

A spectrophotometer is employed to measure the amount of light that a sample absorbs. The instrument operates by passing a beam of light through a sample and measuring the intensity of light reaching a detector.

The beam of light consists of a stream of photons, represented by the purple balls in the simulation shown below. When a photon encounters an analytic molecule (the analyst is the molecule being studied), there is a chance the analyte will absorb the photon. This absorption reduces the number of photons in the beam of light, thereby reducing the intensity of the light beam.

You can visualize this process by running the simulation shown below. Click on the start button to start the simulation and the stop button to stop the simulation the light source is set to emit 10 photons per second Watch the motion of the photons and observe how some of the photons are absorbed as the beam of light passes through the cell containing the sample solution The intensity of the light reaching the detector is less than the intensity emitted by the light source

Experimental Procedure:

The following simulation illustrates the procedures for making spectrophotometric measurements.

1. First of all, the intensity of light (I_0) passing through a blank is measured. The intensity is the number of photons per second .The blank is a solution that is identical to the sample solution except that the blank does not contain the solute that absorbs light. This measurement is necessary because the cell itself scatters some of the light

2. The intensity of light (I) passing through the sample solution is measured (In practice instruments measure the power rather than the intensity of the light. The power is the energy per second, which is the product of the intensity (photons per second) and the energy per photon.

3. The experimental data is used to calculate two quantities:

The transmittance (T) and the absorbance (A)

$T= I/Io$

$A= - \log 10 T$

The transmittance is simply the fraction of light in the original beam that passes through the sample and reaches the detector. The remainder of the light I-T, is the fraction of the light absorbed by the sample (Do not confuse the transmittance with the temperature, which often is given the symbol T)

In most applications one wishes to relate the amount of light absorbed to the concentration of the absorbing molecule. It turns out that the absorbance rather than the transmittance is most useful for this purpose. If no light is absorbed the absorbance is 0 (100% transmittance). Each unit in absorbance corresponds with an order of magnitude in the fraction of light transmitted. For A= I, 10 % of the light is transmitted (T=0.10) and 90 % is absorbed by the sample. For A=2, 1% of the light is transmitted and 99 % is absorbed. For A=3, 0.1 % of the light is transmitted and 99.9% is absorbed.

- Using the simulation below, perform the following steps:
- Measure the intensity of light passing through the blank
- Measure the intensity of light passing through the sample
- Calculate the transmittance
- Calculate the absorbance

Note: For each measurement, run the simulation long enough to detect at least 1000 photons. There is substantial random error in the intensity and the more photons that are counted, the lower the relative uncertainty in the results.

Precautions: Some simple precautions in the use of the spectrophotometer include:

- Allowing the lamps and electronics to warm up
- Using the correct wavelength
- Wiping fingerprints and spilt sample off the outside of the cuvette before measuring
- Carrying out the set-up procedure in the correct order

- Performing calibration checks after set up
- Closing the door to the cuvette compartment before reading the result
- Cleaning up any spills inside the cuvette compartment
- Ensuring that %T or transmission is used as appropriate.

pH Meter:

Measuring pH is about measuring the acidity (pH 0-7) or alkalinity (pH 7-14) of an aqueous sample. The instrument used to perform this function in the laboratory is the pH meter. The main components of a typical meter are:

- Electronic control unit
- Electrode (two electrodes can be used but one is more usual)
- Temperature probe

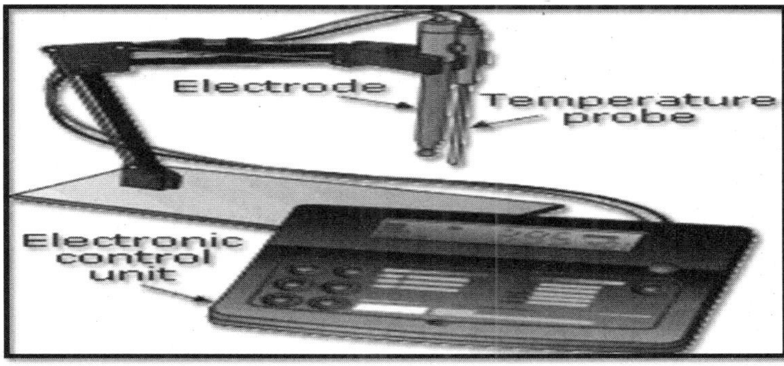

The electrode responds to the concentration of hydrogen ions in the sample solution to produce a chemical signal. This signal is then converted into pH values which are displayed on the control unit.

Variations in temperature can cause pH to change (how much depends on the temperature as well as the nature of the samples). To measure accurately, most pH meters automatically compensate for temperature. For this to occur the control unit needs to know the sample's temperature - this is achieved through the meter's temperature probe, which is immersed in the sample along with the electrode.

When pH meters malfunction, it can often be due to a problem with the electrode. Electrodes are fragile and should be handled and maintained with care. Dirty electrodes or those with scratches or cracks can produce unreliable results. Never leave electrodes to dry out - they should be left immersed in buffer solution between the measuring of samples and during long term storage (never store electrodes in deionised or distilled water). Some electrode types need to be kept filled with an electrolyte solution through a hole near the top of the electrode.

When measuring the pH of a sample/buffer carefully swirl the sample cup around the electrode for a few moments, to ensure the meter responds rapidly and accurately. Always rinse the electrode with distilled water after use and between samples/buffers.

Like any laboratory instrument, pH meters need to be calibrated so that they provide accurate readings. Buffer solutions, which can be made up or purchased for use 'as is', are used for this purpose. These solutions have stable, accurately known pH values, the most common buffers being 4, 7 and 10 because pH varies slightly with temperature, the pH of the buffer solution may not be exactly 4, 7 or 10 (e.g. the pH of buffer solution pH 4 has a pH of 4.0 at 25°C, 4.02 at 30°C). They have a certain shelf life after which they must be replaced. This information is supplied with the buffers.

One point calibrations can be performed (where only one buffer is used) but two point calibrations (using two buffers) give better accuracy overall. Depending on the samples, calibrate the meter initially using the pH 4 or 10 buffer then which calibrate with the pH 7 buffer. You would use the pH 4 buffer when the pH is expected to be less than 7 and the pH 10 buffer when the pH is expected to be greater than 7. In this way, you know that the instrument is adjusted best to cover the pH range of the samples you are testing. If the pH is around 7, either pair of buffers would be satisfactory as would pH 7 buffer if used by itself.

MICROSCOPY

Basic Principles:

Microscopes allow biologists to see single cells and even the organelles that make up their cellular machinery. In some ways, the complex and expensive microscopes they use are like a simple magnifying glass; both tools magnify things by looking at them through pieces of glass (lens). The main difference between them is that while a magnifying glass is made of just a simple lens, today's microscopes are made of increasingly complex sets of lens, and light emission sources. These lens depend on technologies that are at the cutting edge of our understanding of light, which allow those lens to manipulate what we see and don't see.

The basic principle of compound microscopy is that by stacking lens together, it is possible to see increasingly smaller objects. This is similar to stacking magnifying glasses on top of each other, but involves much more complicated optics.

Unfortunately, these are limitations to this system based on the physical properties of light. In order for an image to be in there must be enough light entering the eye, and it must be possible to distinguish between different parts of the image. Over the past 100 years, three developments have made it possible to see different aspects of a biological sample.

How to use a microscope:

- When moving your microscope, always carry it with both hands. Grasp the arm with one hand and place the other hand under the base for support.
- Turn the revolving nosepiece so that the lowest power objective lens is clicked into position (This is also the shortest objective lens).

- The microscope slide should be prepared with a cover slip or cover glass over the specimen. This will help protect the objective lenses if it touch the slide. Place the microscope slide on the stage and fasten it with the stage clips. It can push down on the back end of the stage clip to open it.
- Look at the objective lens and the stage from the side and turn the coarse focus knob so that the objective lens moves downward on the stage, if it moves goes upward). Move it as far as it will go without touching the slide
- Now, look through the eyepiece and adjust the illuminator (or mirror) and diaphragm for the greatest amount of light.
- Slowly turn the coarse adjustment so that the objective lens goes up (away from the slide) continue until the image comes into focus. Use the fine adjustment, if available, for fine focusing .if you have a microscope with a moving stage, then turn the coarse knob so the stage moves downward or away from the objective lens.
- Move the microscope slide around so that the image is in the center of field of view and readjust the mirror, illuminator or diaphragm for the clearest image.
- Now, you should be able to change to the next objective lenses with only minimal use of the focusing adjustment .Use the fine adjustment, if available, if you cannot focus on your specimen, repeat steps 4 through 7 with the higher power objective lens in place, Don't allow the objective lens to touch the slide
- The proper way to use a monocular microscope is to look through the eyepiece with one eye and keep the other eye open (this helps avoid eye strain). If you have to close one eye when looking into the microscope, it's ok. Remember, everything is upside down and backwards. When you move the slide to the right, the images goes to the left.
- Do not touch the glass part of the lenses with your fingers. Use only special lens paper to clean the lenses.
- When finished, raise the tube (or lower the stage), click the low power lens into position and remove the slide.
- Always keep your microscope covered when not in use.
- Remember microscopes are expensive scientific instruments. Handle them properly and carefully and they will last for many years.
- Precautions while using microscope:
- Always set the microscope mirror and diaphragm before placing the slide.
- Focus the slide first under low power; then observe it under high power.
- Focus the slide first by using coarse adjuster and then focus the slide by using fine adjustment knob.
- While focusing at high power, only use fine adjustment. Do not use coarse adjustment knob.

PHASE CONTRAST MICROSCOPY:

Phase contrast microscopy is an optical microscopy technique that converts phase shifts in light passing through a transparent specimen to brightness changes in the image. Phase shifts themselves are invisible, but become visible when shown as brightness variations. When light waves travels through a medium other than vacuum, interaction with the medium causes the wave amplitude and phase to change in a manner dependent on properties of the medium. Changes in amplitude (brightness) arise from the scattering and absorption of light, which is often wavelength dependent and may give rise to colors. Photographic equipment and the human eye are only sensitive to amplitude variations. Without special arrangements, phase changes are therefore invisible. Yet, phase changes often carry important information.

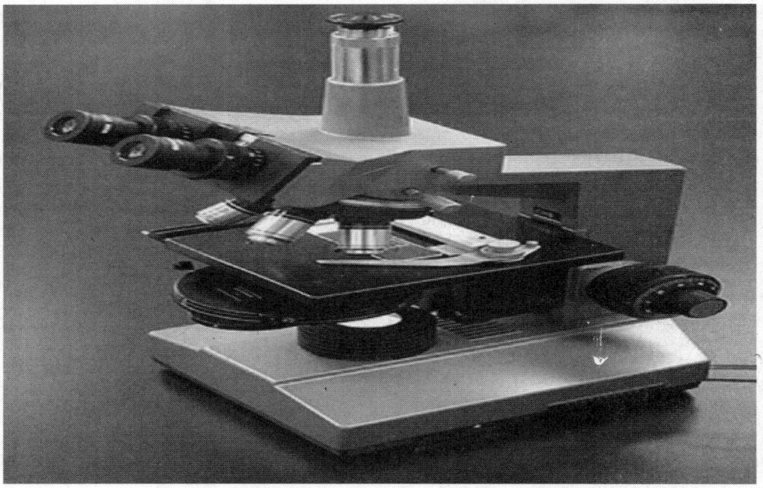

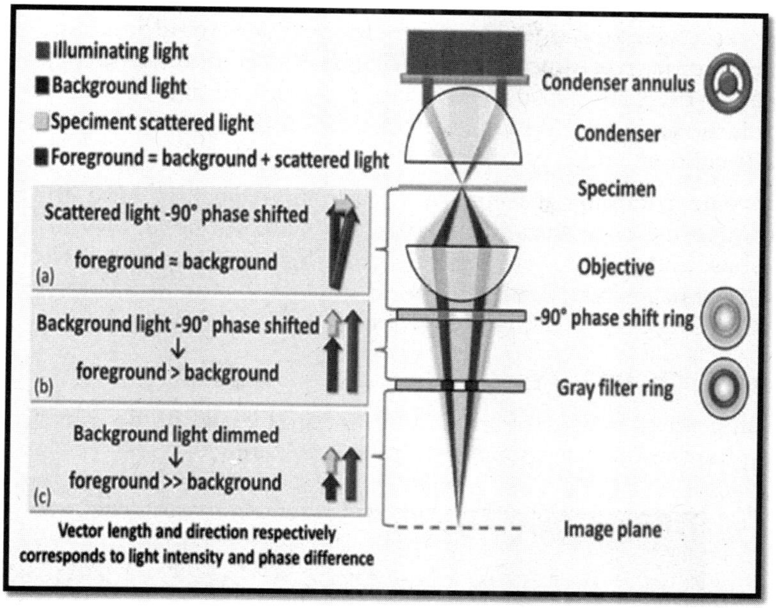

Phase contrast microscopy is particularly important in biology. It reveals many cellular structures that are not visible with a simpler bright field microscope. These structures were made visible to earlier microscopists by staining. This required additional preparation which killed the cells. The phase contrast microscope made it possible for biologists to study living cells and how they proliferate through cell division. After its invention in the early 1930. Phase contrast microscopy proved to be such advancement in microscopy that it inventor Frits Zernike was awarded the Nobel Prize.

Principles

The basic principle to make phase changes visible in phase contrast microscopy is to separate the illuminating background light from the specimen scattered light, which make up the foreground details, and to manipulate these differently.

The ring shaped illuminating light (green) that passes the condenser annulus is focused on the specimen by the condenser. Some of the illuminating light is scattered by the specimen (yellow). The remaining light is unaffected by the specimen and forms the background light (red). When observing unstained biological specimen, the scattered light is weak and typically phase shifted by -90°-relative to the background light. This leads to that the foreground (blue vector) and the background (red vector) nearly has the same intensity, resulting in a low image contrast (a).

In a phase contrast microscope, the image contrast is improved in two steps. The background light is phase shifted 90° by passing it through a phase shift ring. This eliminates the phase difference between the background and the scattered light, leading to an increased intensity difference between foreground and background (b). To further increase contrast, the background is dimmed by a gray filter ring (c). Some of the scattered light will be phase shifted and dimmed by the rings. However, the background light is affected to a much greater extent, which creates the phase contrast effect.

The above describes negative phase contrast. In its positive form, the background light is instead phase shifted by +90°. The background light will thus be 180° out of phase relative to the scattered light. This results in that the scattered light will be subtracted from the background light in (b) to form an image where the foreground is darker than the background.

CHROMATOGRAPHY:

Chromatographic methods can be differentiated based on the physical means of bringing the stationary and mobile phases into contact–

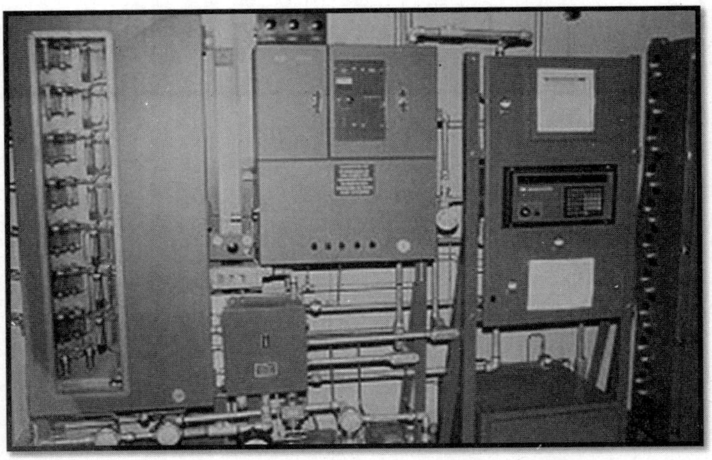

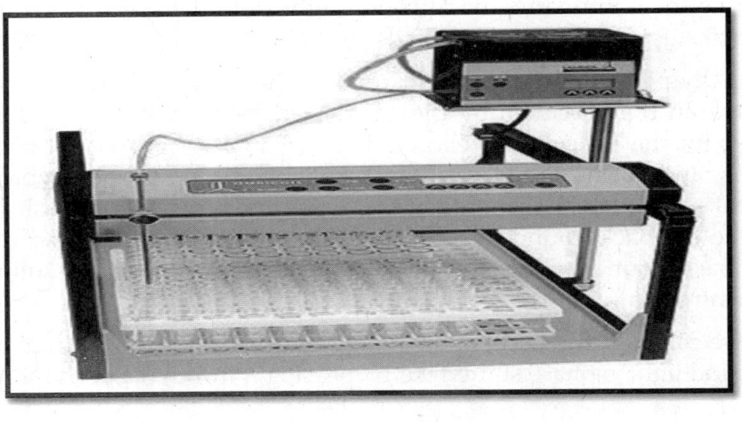

Automated fraction collector and sampler for chromatographic techniques

Column Chromatography: the stationary phase is held in a narrow tube through which the mobile phase is forced either by pressure or by gravity. Examples include:

- Simple column chromatography
- High pressure liquid chromatography (HPLC)
- Gas chromatography (GC).

Planar Chromatography: the stationary phase is supported on a flat plate or in the fibers of a paper. Here the mobile phase moves through the stationary phase by capillary action or by gravity. Examples include:

- Paper chromatography
- Thin layer chromatography (TLC).

Column chromatography: can be further differentiated based on the types of stationary and mobile phases and the kinds of equilibria involved in solute transfer between the phases.There are two broad categories in this classification scheme:

- Liquid chromatography (for instance simple column or HPLC)
- Gas Chromatography (GC).

General Classification	Specific Method Name	Stationary Phase	Type of Equilibrium
Liquid Chromatography (LC) Mobile Phase = liquid	Liquid-liquid or partition	Liquid adsorbed on a solid	Partition between immiscible liquids
	Liquid-bonded phase	Organic species bonded to a solid surface	Partition between liquid and a bonded surface
	Liquid-solid or adsorption	Solid	Adsorption
	Ion exchange	Ion-exchange resin	Ion exchange
	Size exclusion	Liquid in interstices of a polymeric solid	Partition/sieving
Gas Chromatography (GC) Mobile Phase = gas	Gas-liquid	Liquid adsorbed on a solid	Partition between gas and liquid
	Gas-bonded phase	Organic species bonded to a solid surface	Partition between liquid and bonded surface
	Gas-solid	Solid	Adsorption

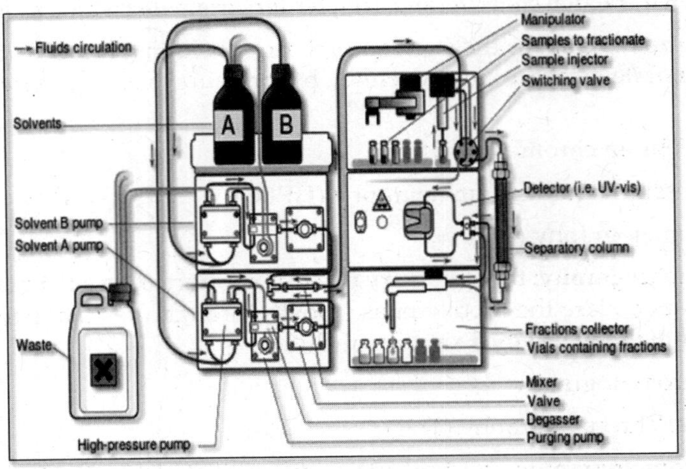

Liquid Chromatography (LC)

Sample preparation:

Sample preparation procedures commonly used with chromatographic and electrophoresis techniques are briefly introduced.

Dissolution:

May be a simple dissolution of the entire sample into a simple solvent (water, detergent solution, ethanol) or may involve extraction by direct separation of the main components by dissolution into a suitable solvent leaving behind insoluble components.

Solvents may be complex such as the sample buffer used for protein SDS-PAGE (sodium dodecyl sulfate polyacrylamide gel electrophoresis), containing the following:

0.01 M Tris HCl, pH 8.0, 0.001 M EDTA, 0.1% SDS

In some cases the solvent may need to be removed prior to analysis.

Concentration:

A variation on dissolution is the precipitation of the sample in a solvent to concentrate the sample. For example, protein samples dissolved in an aqueous salts/detergent buffer may be precipitated in ice-cold acetone. The proteins precipitate out leaving the water and some of the other components in the acetone. The protein pellet is re-dissolved in a smaller volume of buffer.

Dialysis may also be used to concentrate a sample and to change the buffer in which the sample is dissolved. Choice of an appropriate dialysis buffer containing appropriate buffer components and high osmotic strength will cause the water and salts in the solution inside the dialysis bag to migrate out of the bag leading to a concentration of the sample.

Liquid-Liquid Extraction:

A liquid sample (remember that this is also a solvent) is intimately mixed with an immiscible solvent to selectively transfer the analyte from the sample solvent into the immiscible solvent. The transfer occurs as a result of the greater solubility of the analyte in the immiscible solvent. The sample matrix is concentrated in the original sample solvent and the analyte in the immiscible solvent. After extraction, the immiscible solvent can be separated from the analyte (eg by evaporation) in readiness for analysis or for further processing.

This method requires a good knowledge of chemistry and the sample components need to be known in advance. This procedure facilitates purification and/or concentration of the sample.

The first essential step in DNA analysis is to separate the DNA from the cellular components (matrix) in which it is contained. One of the most commonly used methods is to extract the DNA in a chloroform/phenol liquid-liquid mix.

Solid Phase Extraction and Affinity Chromatography:

Adsorption of molecules of interest onto a solid, that is a stationary phase, of appropriate material in a column (column chromatography) or a planar surface (thin layer chromatography) are often used to clean up samples prior to further chromatographic or electrophoretic analysis. The molecules of interest are eluted (i.e. flushed off) from the solid phase using an appropriate solvent.

In affinity chromatography, a specific ligand is attached to the chromatographic bed material. The sample is passed through the bed and the molecules of interest bind to the bed via the specific ligand. These molecules of interest can then be eluted from the column by changing the buffer conditions.

Filtration and Centrifugation:

Chromatographic techniques such as HPLC and GC rely on fine tubing and columns with fine orifices. The presence of a precipitate or un-dissolved matter from some other source in the test sample may block or retard migration of the sample along the column. Removal of the un-dissolved material by filtering or centrifuging the sample can overcome this problem.

Introduction:

The chromatography and how things work so what is chromatography used for and what can it do?

Qualitative analysis:

The chromatogram of a run provides one piece of information about each species in a sample, the retention time or position, but this information is important and can be used to identify multiple species in a mixture.

With experience it is a relatively simple task to identify species by their position but in initial work the identity of each separated species on the chromatogram should be confirmed by analytical methods such as nuclear magnetic resonance or mass spectroscopy. That is, a chromatogram does not positively identify a species in a mixture.

Of course a chromatogram also provides information of the absence of a species from a sample. For example, lack of steroids from an Olympic weightlifter's urine sample is seen as a great benefit to the career of the weightlifter!

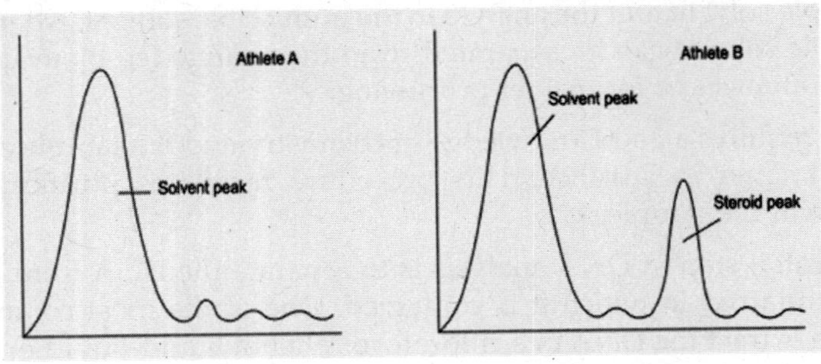

Quantitative analysis:

Chromatography also provides some important qualitative information about the separated species present in a mixture. For column chromatography, comparison of the height of a peak or the area under a peak can be compared with standards to determine the amount of that species present. For planar chromatography the area covered by the spot of the separated species can be measured and again compared to a standard to arrive at an estimation of the amount of the species present.

Standards are usually run separately and it is crucial to ensure conditions do not change between runs. In some circumstances standards may be run with the sample and therefore conditions should not change during the run that will affect the peak heights or widths.

If the run is set up correctly these measured parameters vary linearly with concentration.

Analyses based on peak height:

This is measured as shown in the following diagram. Essentially the base lines on either side of the peak are connected with a straight line and the perpendicular distance to the peak is measured. Peak broadening can reduce the accuracy of this approach.

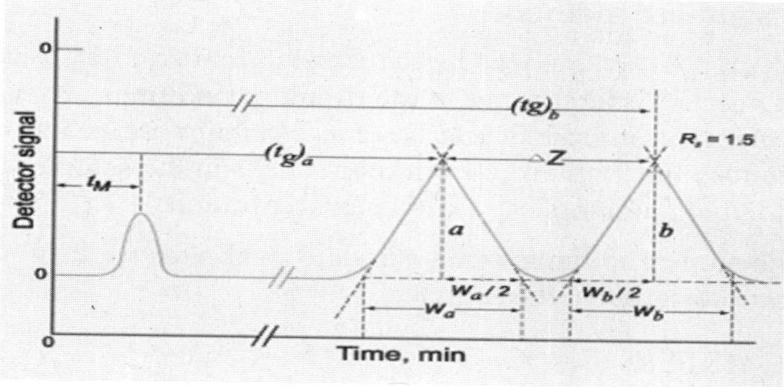

Analyses based on peak areas:

Peak areas are independent of peak broadening effects and are therefore a better measure of concentration.

Modern chromatography equipment have digital integrators that will automatically measure areas under the peaks but if these are lacking then manual methods must be used.

These include:

- Cutting out the peak from the chromatogram, weighing on a balance and comparing the mass to a standard area of say 100 cm^2
- For high, narrow, symmetric peaks assume they are triangles and multiply the height of the peak by one half of the width of the base of the peak.

Calibration and standards:

The easiest approach is to prepare a series of standard solutions that approximate the composition of the unknown. The chromatograms of the standards are used to plot the peak heights or areas against concentration to obtain a standard curve. This should be a straight line that passes through the origin.

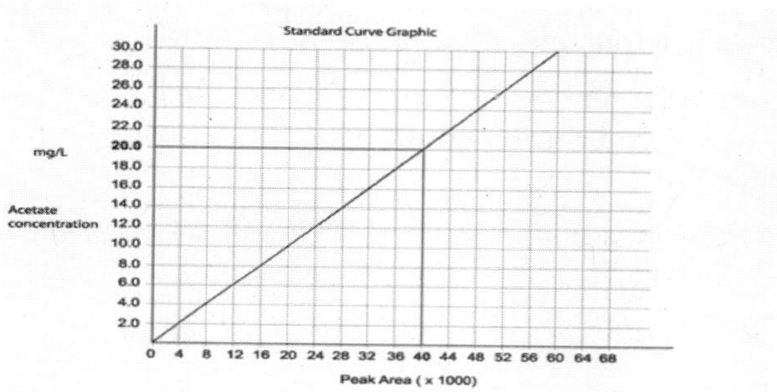

The standard curve is then used to determine the concentration of the unknown sample. Of course the run conditions must be as close to identical as possible and frequent re-standardization is required for accuracy.

The internal standard method:

The greatest accuracy is obtained by the use of an internal standard as uncertainties due to sample injection and changes to the running conditions are eliminated. In this case the internal standard should be of a different species and run close to the peak of interest, but well away from other peaks in the sample. The internal standard is added to the sample at a known concentration.

The ratio of the analyte to the internal standard peak area (or height) is used to calculate the concentration of the analyte.

HEMOCYTOMETER:

The simplest, most convenient and cheapest means of accurately determining the numbers of cells in a sample is to use a Haemocytometer and a microscope. A Haemocytometer is a specialized slide that has a counting chamber with a known volume of liquid.

- The Haemocytometer consists of a heavy glass slide with two counting chambers, each of which is divided into nine large 1 mm squares, on an etched and silvered surface separated by a trough.
- A cover slip sits on top of the raised supports of the 'H' shaped toughs enclosing both chambers. There is a 'V' or notch at either end where the cell suspension is loaded into the Haemocytometer. When loaded with the cell suspension it contains a defined volume of liquid.
- The engraved grid on the surface of the counting chamber ensures that the number of particles in a defined volume of liquid is counted.
- The Haemocytometer is placed on the microscope stage and the cell suspension is counted.

Principles

- This is one of the two chambers in the Haemocytometer.
- The entire chamber has its nine 1.0 mm × 1.0 mm large squares separated from one another by the triple lines. Area of each is 1 mm^2.
- Within each of the larger corner 1 mm^2 squares are 16 small squares that are there to help orient you during counting, to help avoid counting a given cell more than once.
- The central 1 mm^2 area is divided into 25 small squares, each 0.04 mm^2 and each of these is marked into a further 16 tiny squares.

Dimensions	Area	Volume at 0.1 mm depth
1 x 1 mm	1 mm^2	100 nl
0.25 x 0.25 mm	0.0625 mm^2	6.25 nl
0.25 x 0.20 mm	0.05 mm^2	5 nl
0.20 x 0.20 mm	0.04 mm^2	4 nl
0.05 x 0.05 mm	0.0025 mm^2	0.25 nl

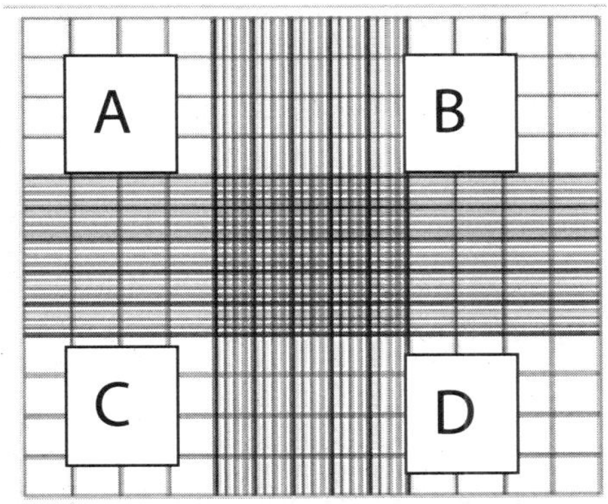

Procedure:

Prepare a cell/spore/conidial suspension by filtering the fungal culture suspension through muslin cloth to get rid of mycelium.

- Prepare a hemocytometer for use by carefully cleaning all surfaces of the hemocytometerand coverslip.
- Center the cover slip on the hemocytometer.
- Pipette approximately 50 µl (this volume will vary slightly with the brand of hemocytometer) of the spore suspension into one of the two counting chambers using a clean pipette tip.
- Fill the chambers slowly and steadily avoiding injecting bubbles into the chambers.
- Count all the spores in each of the four corner squares labeled A to D.

Determine the spores count:

- Calculated the total spores counted in the all four corner squares.
- Calculated the total spores count using the following equation:

Spores/ml = (n) X 104

Where; n = The average cell count of the four corner squares counted

E.g. If the calculated average (n) of spores in the four 1 mm corner squares of the hemocytometer is 30 Spores/ml = 30 X 104 = 300,000 Spores/ml

or

$$\text{Spores/ml} = \frac{\text{Average no. of spores in one block} \times \text{No. of central small square} \times \text{Dilution factor}}{\text{Volume in Central Square in ml}}$$

Eg.

For 5 blocks

-In case of 10ml volume taken for each block so that 5 blocks volume required = 50ml = 40 + 45 + 55 + 60 + 50 = 250/5 = 50

Cfu/ml = 50 x 25 x 104 =1250 x 104 or 1.25 x 107

Viable (coloured) + non-viable (Blue)

Requirements:

- The original suspension must be mixed thoroughly before taking a sample. This ensures the sample is representative, and not just an artifact of the particular region of the original mixture it was drawn from.
- An appropriate dilution of the mixture with regard to the number of cells to be counted should be used. If the sample is not diluted enough, the cells will be too crowded and difficult to count. If it is too dilute, the sample size will not be enough to make strong inferences about the concentration in the original mixture.
- By performing a redundant test on a second chamber, the results can be compared. If they differ greatly, the method of taking the sample may be unreliable (e.g., the original mixture is not mixed thoroughly).

CAMERA LUCIDA:

A camera lucida is an optical device used as a drawing aid by artists. The camera lucida performs an optical super imposition of the subject being viewed upon the surface upon which the artist is drawing. The artist sees both scene and drawing surface simultaneously, as in a photographic double exposure. This allows the artist to duplicate key points of the scene on the drawing surface, thus aiding in the accurate rendering of perspective.

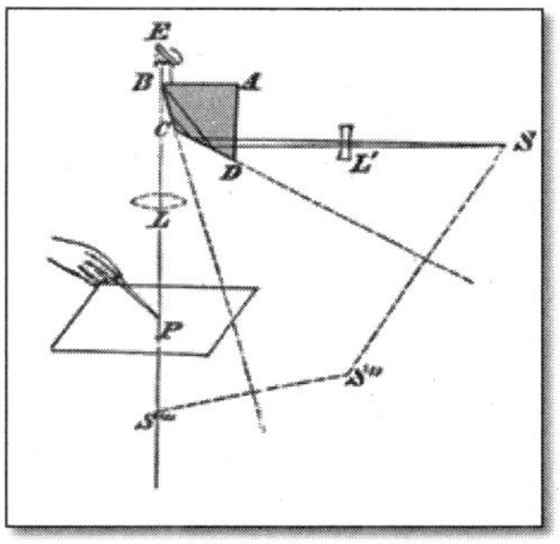

The name "Camera Lucida" (Latin for "light chamber") is obviously intended to recall the much older drawing aid, the camera obscura (Latin for "dark chamber"). There is no optical similarity between the devices. The camera lucida is a light, portable device that does not require special lighting conditions. No image is projected by the camera lucida.

Use of camera lucida :

A camera Lucida reflects a subject so that it appears to be on your piece of paper, enabling you to simply trace it. The following is based on using a camera lucida made by The Camera Lucida Company, but they all work similarly.

Setting of camera lucida:

Set up the drawing board up at an angle of 40 degrees; putting it on your lap and resting it against the edge of a table works well. Put a piece of paper on the board, up to A3 in size. Swing up the arm with the 'viewing lens' up, twist the 'lens' so that the small eye hole is at the top. When you look through this, you should be able to see the whole piece of paper and the scene as if reflected on it.

Subject on the paper:

Check the position of the camera's viewer. Are you looking down towards the paper? If so, it's a question of getting the balance of light between your subject and the paper right. Place a piece of black paper on the drawing board; if you can now see the subject, you should light it more. If you can't see the piece of paper because the subject's too strong, use a lamp to throw a bit more light on your paper. At times you'll find there are parts that are too light or too dark to see detail; you could fiddle with getting the light balance just right, or simply use your other eye or look up at the actual scene to see what's there.

If white paper is used with the camera lucida, the superimposition of the paper with the scene tends to wash out the scene, making it difficult to view. When working with a camera lucida it is often beneficial to use black paper and to draw with a white pencil.

19 PREPARATIONS OF BUFFER AND REAGENTS

Given below are the methods for preparing buffers & reagents which are generally used in a molecular biology lab. The solutions which are specific for any experiment are given under the respective headings:

30% Acrylamide

Acrylamide	30.0 g
N-N methylene bisacrylamide	0.8 g

Dissolve in water and make up vol. To 100 ml. Store in a dark bottle at 4°C for upto 2 weeks.Acrylamide is neurotoxic. Wear gloves while handling.

0.1% M Adenosine Triphosphate (ATP): Dissolve 60 mg of ATP in 0.8 ml water. Adjust the pH to 7.0 with 0.1 NaoH. Make upto 1 ml. Dispence into aliquotes. Store at -70°C.

10% Ammonium persulphate (APS): Dissolve 1 g of APS in sterile distilled water & make upto 10 ml .Store at 4°C. It is preferable to prepare freshly.

Antibiotics: Dissolve 5 mg of ampicillin/kanamycin in 10 ml of sterile double distilled water. Filters sterilize and dispense into 200 µl aliquots and store at -20°C.

Antigen extraction buffer (TBS + 50 mM DIECA) (for DIBA): Add 11.25 g diethyl dithiocarbamate (DIECA) to 1L TBS.

Antibody and enzyme-conjugate diluents/buffer (for ELISA): Same as TBS-SDM.

Antibody diluents/buffer (for ELISA): Same as PBS-TPO.

Blocking solution (for ELISA): Add 5 g Bovine Serum Albumin (BSA)/Spray dried milk (SDM) to 1l PBS-T.

Stock solution

NaoH	12 g
Distilled water	to make 1 L.

All buffers contain 0.02% sodium azide as a preservative.

Blocking solution (TBS+SDM): Add 5 g spray dried milk (SDM) to 100 ml TBS.

Bind –saline

10% Acetic acid	375 μl
Bind silane	37.5 μl

Make upto 12.5 ml with absolute ethanol.

100 mM $Cacl_2$: Add 1.47 g of Cacl2.2H2O in H2O. Make upto 100 ml Sterlise by autoclaving.

Coating buffer (Carbonate buffer, pH 9.6)

Na_2CO_3	1.59 g
$NaHCO_3$	2.93 g
Distilled water	to make 1 L

Molecular weight marker: Commercial molecular weight markers are normally prepared from digestion of different plasmids to produce discrete fragments.

DNA (ladder)	100 μl
50 X TAE	5 μl
6 X Gel loading dye	42 μl
SDW	103 μl
Total	250 μl

The marker is ready to be loaded.

Denhardt's Solution (50 X): Dissolve 5 g of Ficoll, 5 g polyvinyl pyrrolidone and 5 g of BSA in H_2O & make upto 500 ml. Filter & dispence into 25 ml aliquotes. Store at -20°C.

Destaining solution:

Methanol	50 ml
Glacial acetic acid	75 ml
Distilled water	1000 ml

Denaturing solution (1.5 M Nacl, 0.5 N NaoH)

5M Nacl	30 ml
10 N NaoH	5 ml
SDW	65 ml
Total	100 ml

0.5 M EDTA (pH 8): Add 186.1 g of disodium ethylene demine tetraacetate.$2H_2O$ to 800 ml of H_2O .Stir vigorously on a magnetic stirrer. Adjust the pH to 8.0 with NaoH. Make upto 1 L. Dispense into aliquotes & sterilize by autoclaving.

Ethidium Bromide (10 mg/ml) : Add 1 g ethidium bromide to 100 ml of H_2O. Stir on a magnetic stirrer for several hrs. to ensure that the dye has dissolved .Wrap the container in aluminium foil or transfer to a dark bottle & store at 4°C.

Caution: Ethidium bromide is a mutagen & toxic. Wear gloves when working with ethidium bromide solutions & a mask when weighing it out.

Equilibrium of Phenol: Melt 500 g phenol in a water bath, kept at 65°C. Add 100 ml of 100 mM Tris HCl (pH 8) & shake to emulsify. Let phases separate. Repeat this step until the aqueous layer shows a pH between 7 to 7.5. Store the phenol under 50 ml 0f 10 ml Tris HCL pH 7.6 in an amber coloured bottle.

Enzyme conjugate diluent/buffer (PBS T polyvinyl-pyrrolidone ovalbumin,PBS –TPO) (for ELISA) : Add 20 g polyvinylpyrorolidone (PVP, MW 44000) & 2 g egg ovalbumin to1 L PBS-T.

Extraction buffer (Total nucleic acid)

2% w/v CTAB

1.4 M NaCl
50 Mm EDTA pH 8.0
100 mM Tris –Hcl pH 8.0
0.2% β-mercaptoethanol (Add just before use)

Fixative solution

Methanol 400 ml

Glacial acid	70 ml
Distilled water	1000 ml

Fixing solution (pH 7.5)(for DIBA)

10 mM Tris	1.21 g
1 mM EDTAo.	29 g

Adjust the pH to7.5 with 1 N HCL and make upto 1L. All buffers contain 0.02% sodium azide as a preservative.

IPTG (isopropyl thio β-galactoside): Dissolve 2.0 g of IPTG in 8 ml of distilled water. Make up to 10 ml. Filter sterilize through a 0.22 μ disposable filter. Dispense into aliquots. Store at -20°C.

L A medium (Luria agar medium): 950 ml of deionized water, add 10 g of bactotryptone, 5 g of yeast extract & 10 g NaCl & 15 g agar. Boil to dissolve content, adjust pH to 6.8 with 5 N NaOH and make up volume up to 1l with deionized water. Sterilize by autoclaving.

Luria-Bertani broth: It is same as above but prepared without agar.

Loading dye (6X)

50 mM	Tris –HCL, pH 8.0
5 mM	EDTA
25%	Glycerol
0.2%	Bromophenol blue
0.2%	Xylene Cyanol FF

This loading dye is used for Non-denaturing ds DNA

6 X loading dye

1% Bromophenol blue	:	200 μl
Glycerol	:	200 μl
10% SDS	:	60 μl
0.5 M EDTA	:	50 μl
10 X TAE	:	60 μl
Water	:	30 μl

$MgCl_2$: Add 2.03 g of $MgCl_2$. $6H_2O$ in H_2O & make upto 100 ml. Sterilise by autoclaving.

5M NaCl : Dissolve 292.2 g of NaCl in H_2O and make up to 1 litre. Dispense into aliquots. Sterilize by autoclaving.

Nutrient agar (NA)

Beef extract	:	5.0 g
Peptone	:	10 g
NaCl	:	5.0 g
Agar	:	20.0 g
Distilled Water	:	1000 ml
pH	:	7.0

Nutrient broth (NB): Same as NA, But without agar.

Neutralizing solution [1 M Tris- HCl (pH 7.4) 1.5 M NaCl]

2M Tris	:	50 ml
5M NaCl	:	30 ml
SDW	:	100 ml

Plasmid isolation buffer (Modified alkaline lysis method)

Solution I	:	25 mM Tris HCl (pH 8.0)
(Resuspension buffer)	:	50 mM Glucose
	:	10 mM EDTA

Solution II (Lysis buffer)	:	0.2 N NaOH
	:	1% SDS
Solution III (Neutralization buffer)	:	3 M sodium acetate(pH 4.8)

Prehybridisation solution

50 X denhardt's solution	:	1 ml
20 X SSC	:	3 ml
10% SDS	:	0.3 ml
SDW	:	5.7 ml
Final volume	:	10 ml

Add large molecular weight DNA like calf thyms DNA or salmen sperm DNA at the rate of 10 ug/ml. Yeast RNA can also be added.

Phosphate buffer-saline, PBS, pH 7.4

NaCl	:	8.00 g
$Na_2HPO_4.2H_2O$	:	1.44 g
or		
$Na_2 HPO_4 .12 H_2O$	:	2.90 g
or		
$Na_2 HPO_4$	:	1.50 g
KH_2PO_4	:	0.20 g
KCl	:	0.20 g

Distilled water to make 1 litre.

Resolving gel solution

Acrylamide stock solution	:	19.8 ml
Tris-SDS stock solution (pH 8.8)	:	30 ml
Distilled water	:	to 60 ml
TEMED	:	15 μl
10% APS	:	1 ml

10 % Sodium dodecyl Sulfate (SDS): Dissolve 100 g of SDS in H_2O and make up to 1 liter. Dispense into aliquots.

20X SSC: Dissolve 175.3 g of NaCl and 88.2 g of sodium citrate in H_2O and make up to 1 liter. Dispense into aliquots and sterilize by autoclaving.

3M Sodium acetate (pH 4.8 and 5.2): Dissolve 408.1 g of sodium acetate $3H_2O$ in 800 ml of distilled water, Adjust the pH to 4.8 to 5.2 with glacial acetic acid. Make up to 1 liter. Dispense into aliquots and sterilize by autoclaving.

Staining solution

Coomasive Brilliant Blue R 250 (Dissolve in 227 ml methanol)	:	1.15 g
Glacial acetic acid	:	46 ml
Distilled Water	:	50 ml

Substrate buffer (diethanolamine buffer pH 9.8) (for ELISA)

Diethanolamine	:	97 ml
Distilled water	:	800 ml

Adjust the pH to 9.8 with 1 N HCl and make up the volume up to 1litre.

Substrate Buffer (pH 9.5) for DIBA

Solution A

Nitro Blue Tetrazolium (NBT)	75 g
Dimethyl formamyde (DMF)	1 ml

Solution B

Bromo chloro indolyl phosphate (BCIP)	50 mg
DMF	1 ml

Store solution A and B refrigerated in amber coloured bottle. Add NBT to 0.33 mg/ml and BCIP to 0.175 mg/ml to the substrate buffer just before use.

Sample Buffer

Tris-SDS Stock pH 6.8	:	25 ml
SDS	:	2.0 g
Glycerol	:	10 ml
2-mercaptothanol	:	10 ml
1% bromo phenol blue	:	0.1 ml
Distilled water	:	100 ml

Stacking gel (for two 1.5 mm gel, 3%)

Acrylamide stock solution	:	2 ml
Tris-SDS stock sol (pH6.8)	:	10.0 ml
Dt water	:	60 ml
TEMED	:	70 μl
10% APS	:	1 ml

1M Tris-HCl: Dissolve 121.1 g of Tris hydroxylmethane aminomethane in 800 ml distilled water. Adjust the pH to 8.0 with hydrochloric acid. Make up to 1 liter. Dispense into aliquots and sterilize by autoclaving.

Tris-buffer saline, TBS pH 7.5

0.02 M Tris	:	4.84 g
0.5 M NaCl	:	58.48 g

Adjust the pH to 7.5 with 1N HCl and make up to 2 liter. This is used as wash solution

Tris-Glycine-SDS electrode buffer, pH 8.3

Tris	:	15.15 g
Glycine	:	72.0 g
SDS	:	5.0 g

Disslove in water and make up the volume to 5000 ml.

Tris -SDS stock solution pH 6.8

Tris	:	39.4 g
SDS	:	2 g

Dissolve in 900 ml distilled water. Adjust the pH to 6.8 with HCl and make up the volume to 1000 ml

Tris- SDS stock solution pH 8.8

Tris	:	118.2 g
SDS	:	2 g

Dissolve in 900 ml distilled water. Adjust the pH to 8.8 with HCl and make up the volume to 1000 ml

Transfer buffer (10 X SSC)

20 X SSC	50 ml
SDW	50 ml

10 X TBE (500 ml)

Tris	54 g
Boric acid	27.5 g
EDTA (sodium salt)	4.65 g

TE saturated phenol: Chloroform: Isoamyl alchohal: To the equilibriated phenol add equal amount of chloroform: Isoamyl alchohal (24:1). Store at 4°C in an amber coloured bottle under a layer of 10 mM Tris-Cl (pH 7.6).

50 X TAE: Dissolve 242 g of Tris in H_2O. Add 57.1 ml of glacial acetic acid and add 100 ml of 0.5 M EDTA (pH 8.0) Make up to 1 liter. Dispense into aliquots and sterilize by autoclaving.

Wash buffer (0.1%SDS, 2X SSC)

10% SDS	:	2 ml
20 X SSC	:	20 ml
SDW	:	178 ml
Final volume	:	200 ml

Wash buffer (PBS-TWeen, PBS-T) (For ELISA): Add 0.5 ml Tween -20 to 1 liter PBS.

X-gel (5-bromo-4chloro-3 indolyl-B-D- galactopyranoside): dissolve X-gel in dimethyl formamide to make a 20 mg/ml solution and store at -20°C.

20 METHODS OF BIOCHEMICAL ANALYSIS

- ❖ **Estimation of Chlorophyll**
- ❖ **Estimation of Proline**
- ❖ **Estimation of Phenol**
- ❖ **Estimation of Protein**
- ❖ **Estimation of Total sugar (Reducing and Non-Reducing)**
- ❖ **Estimation of Polyphenol Oxidase (PPO)**
- ❖ **Estimation of Catalase (CAT) enzyme**
- ❖ **Estimation of Phenyl-alanine Ammonia Lyase (PAL)**
- ❖ **Estimation of Peroxidase (POD)**
- ❖ **Estimation of Superoxide Dismutase (SOD)**

Estimation of chlorophyll: Chlorophylls are present as green pigments in all green plant tissues. Chlorophyll *a* and *b* occurs in higher plants. Diseases development affects not only the total chlorophyll content but also alters the ration between chlorophyll *a* and *b*. measurement of green pigment may indirectly denote the severity of diseases especially in systemic virus diseases. Chlorophyll estimates may also be required to relate other biochemical changes in the plant tissues. Assays of chlorophyll content provide an efficient measure of chloroplast concentration.

Principle:

The chlorophyll is extracted from the chloroplast by shaking with 80% acetone and the absorbance is measured at 645 and 663 nm for determination of chlorophyll *a* and *b* and total chlorophyll.

Reagent: Acetone (80%, pre-chilled)

Procedure: Total chlorophyll content was estimated by the method (**Witham *et al.* (1971)**).

- ❖ Chlorophyll was extracted from 1 g of the sample using 20 ml of 80% acetone. The supernatant was transferred to a volumetric flask after centrifugation at 5000 rpm for 5 minutes.
- ❖ The extraction was repeated until the residue became colourless.
- ❖ The volume in the flask was made up to 100 ml with 80% acetone.
- ❖ The absorbance of the extract was read in a spectrophotometer at 645 and 663 nm against 80% acetone blank.

- The amount of total chlorophyll in the sample was calculated using the formula.

Chl A = $(12.7 \times A_{663} - 2.69 \times A_{645})$; Chl B= $(22.9 \times A_{645} - 4.68 \times A_{663})$

Total chlorophyll = $20.2\ (A_{645}) + 8.02\ (A_{663}) \times 1000 \times W$ where, V = final volume of the extract: W = fresh weight of the leaves. The values are expressed as mg chlorophyll/g sample.

Estimation of proline: Proline is an heterocyclic amino acid found abundantly in basic proteins. Free proline in plants is playing a role under pathological stress conditions. In diseased plants, the protins is breakdown into amino acid resulted enhanced level of proline and might be related to pathological disorder.

Principle:

Proteins are precipitated as a protein-sulphosalicyclic acid complex during extraction of tissue with sulphosalicylic acid. The extracted proline is made to react with ninhydrin under acidic conditions to form a red colour which is measured calorimetrically at 520 nm.

Reagents:

- Acid ninhydrin reagents,
- 6N Orthophosphoric acid,
- Glacial acetic acid,
- Sulphosalicyclic acid,
- Toluene
- Standard proline solution.

Procedure: Free proline in leaves (mg g^{-1} fresh weight) was extracted and determined by the methods of **Bates et al, (1973)**.

- 100 mg fresh leaf sample was homogenized in 5 ml 3% aqueous sulphosalicylic acid and the supernatant was collected after centrifugation at 5000 rpm for 5 min.
- The volume of supernatant was made to 10 ml with the sulphosalicylic acid.
- The sample extract of 2 ml was taken in one test tube and at the same time

in a series of test tubes 0.2, 0.4, 0.6, 0.8 and 1.0 ml of proline solutions were prepared by dissolving 10 mg of proline in 100 ml of 3% sulphosalicylic acid.

- A tube with 2 ml sulphosalicylic acid was served as blank.
- In all test tubes 2 ml ninhydrin reagent (1.25 g ninhydrin 10 20 ml of 6N orthophosphoric acid) and 2 ml glacial acetic acid were added and kept in boiling water bath at 100°C for 10 min and then cooled in ice bucket to stop the reaction.
- Then, 4 ml toluene was added and mixed thoroughly by vortex mixer for 10 seconds.
- Optical density was measured at 520 nm against a reagent blank using UV visible spectrophotometer.
- The quantity of free proline in 100 mg of leaf was then calculated using the standard curve.

Estimation of phenol: Plants contain a large number of aromatic compounds with hydroxyl groups which are collectively referred to as phenolic or phenols. Infection of plants induces changes in phenolic substances. Varieties differ in their response with regards to phenolic changes, initially an increase in the susceptible and resistant varieties but with the symptoms development. Phenols decrease in susceptible varieties while in the resistant varieties phenols accumulated.

Principle

Phenols react with an oxidizing agents phosphomolybdate in Folin- Ciocalteu reagent under alkaline conditions and result in the formation of a blue coloured complex, the molybdenum blue which is measured at 650 nm calorimetrically **(Bray and Thorpe, 1954).**

Reagents:

- 80% Ethyl alcohol,
- FCR,
- 20% Sodium carbonate
- Standard (100 mg catechol in 100 ml of water) Dilute 10 times for a working standard.

Procedure: Total phenol from plant samples were estimated by using **(Bray and Thorpe, 1954)** method.

- Weigh 500 mg plant leaves samples and homogenate with 10 ml 80% ethanol. Then centrifuge at 10000 rpm for 10 min.
- The supernatant was taken in test tubes and again re-extract the residue with 5 ml of 80% ethanol.
- Pull the supernatant and dry.
- Add 5 ml of distilled water in a dry residue and again add 0.5 ml of FCR in each tube.

- After 3 min. add 2 ml Na_2CO_3 (20%) and mix thoroughly the each test tube and place all tubes in a boiling water bath for 1 min.
- Cool all the tubes with cold water and takes reading at 650 nm in spectrophotometer.
- Prepared a standard curve using different concentrations of catechol. From the standard curve find out the concentration of phenols in the test sample and express as mg phenols/100g material.

Estimation of protein: The **Lowry** reaction for protein estimation is an extension of the biuret method. This method was 10 times more sensitive than biuret method and largely followed to determine the protein content of enzyme extract.

Principle:

Protein reacts with FCR to give a blue coloured complex. The colour formed is due to the reactions of the alkaline copper with the protein as in the biuret test and the reduction of phosphomolybdic – phospotungstic component in the FCR by the amino acids tyrosine and tryptophan present in the protein. The intensity of blue colour is measured at 660 nm under spectrophotometer.

Reagents:

- BSA stock solution (1mg/ml)
- Analytical reagents: Solution (a) 50 ml 2% sodium carbonate mixed with 50 ml of 0.1 N NaOH solutions (0.4 g in 100 ml distilled water.)
- Solution (b) 10 ml of 0.5% copper sulphate solution mixed with 10 ml of 1% sodium potassium tartrate solution.
- Prepare analytical reagents (Solution C) by mixing 2 ml of (b) with 100 ml of (a).
- Folin - Ciocalteau reagent solution (1N) Dilute commercial reagent (2N) with an equal volume of water on the day of use (2 ml of commercial reagent + 2 ml distilled water).
- Working standard solution: Dilute 10 ml of the stock solution to 50 ml with water to obtain 200 μg protein/ ml.
- Stock standard protein solution: 50 mg bovine serum albumin in 50 ml of distilled water.

Procedure:

Extraction of protein from leaf samples:

- 500 mg of leaf sample from plant was homogenate with water or buffer and made upto 1ml solution.
- Centrifugation at 10000 rpm for 20 minutes was done and the supernatant was used for protein estimation.

Estimation of Protein

- Different dilutions of BSA solutions were prepared by mixing stock BSA solution (1 mg/ ml) and water in the test tubes.
- From these different dilutions, 0.2 ml protein solution was pipetted out to different test tubes and 5 ml of alkaline copper sulphate reagent (Solution C) was added and solutions were mixed well.
- This solution is incubated at room temperature for 10 min.
- Then 0.5 ml of reagent Folin-Ciocalteau solution (reagent solutions) was added to each tube and incubate for 30 min. Zero the Spectrophotometer with blank and the optical density, the absorbance was measured at 660 nm.
- The absorbance against protein concentration was plotted to get a standard calibration curve.
- Check the absorbance of unknown sample and determine the concentration of the unknown sample using the standard curve plotted above.
- Standard graph was drawn the amount of protein in the sample was calculated and express the results as mg/g or mg/100g sample or percentage.

Estimation of total sugar (Reducing and Non-Reducing)

a) Estimation of non- reducing sugar

Principle:

Non reducing sugars present in the plant extract are first hydrolyzed with either sulphuric acid or formic acid to reducing sugars. Then, the total non-reducing sugars are estimated by Nelson –Somogyi method.

Reagents:

- 1N H_2SO_4 (4 ml in 150 ml water)
- 1N NaOH (6 g in 150 ml water)
- Methyl red indicator (1 or 2 drop)

Procedure:

- Weigh 100 mg leaf samples and homogenate with water.
- Take 5 ml of 80% ethanol (80 ml ethanol + 20 ml water) in homogenate sample and centrifuge at 10000 rpm for 5 min.
- Take the supernatant in the test tubes and evaporate on water bath at 63°C.
- Added 10 ml of water to dissolved the sugar then pipette out 1 ml of extract and 1 ml of 1N H_2SO_4 in test tubes.
- Heated all the test tubes at 49°C for 30 min.
- Cool the tubes in cold water and add 1 or 2 drops of the methyl red indicator.
- Neutralized the content by adding 1N NaOH drop wise.
- Take the reading at 620 nm in spectrophotometer.

b) Estimation of total soluble sugar:

- To estimate the total soluble sugar, 100 mg of respective leaves were taken and homogenate in 80 percent (v/v) ethanol.
- Homogenate was centrifuged thrice.
- Supernatant fraction thus separated was pooled and final volume was made to 25 ml with 80% ethanol.
- Ethanol soluble sugars were estimated in the supernatant. After fraction of soluble sugar, 0.5 ml of supernatant was taken in separated test tube and dried in oven at 60°C. to the dried material
- 1 ml distilled water was added followed by 5 ml of freshly prepared anthrone reagents which was prepared by dissolving 2 g anthrone in 1000 ml concentration of H_2SO_4.
- Tubes were kept in boiling water bath for 10 minutes then cooled.
- Absorbance of green colour was measured at 620 nm with the help of spectrophotometer. Standard curve was prepared by taking the known amount of glucose.

C) Estimation of reducing sugar: Reducing sugar will be calculated by subtraction of non- reducing sugar from total soluble sugar.

Estimation of Polyphenol Oxidase (PPO): Polyphenol oxidase is one of the enzyme involved in the oxidation of phenolic compounds to brown pigments. In plant it is suggested that the enzyme is involved in resistance to infection. The enzyme catalyses the oxidation of monophenols and o-diphenols.

Principle:

The enzyme activity is measured as rate of increase in absorbance calorimetrically at 410 nm with the oxidation of catechol as the substrate.

Reagents:

- Chilled acetone
- 0.2M potassium phosphate buffer, pH 6.8.
- 0.05M Catechol: Prepared in 0.2M phosphate buffer, pH 6.8.

Procedure:

Enzyme extraction:

- The leaf material was immersed in acetone and crushed and dried then made to powder. 5g of this acetone powder was dissolved in 200 ml of 0.2M potassium phosphate buffer, pH 6.8 and stirred well for 30 min.
- The above mixture was then centrifuged at 11000g for 20 min at 2°C. The supernatant dialyzed against 0.2 M phosphate buffer for 2 days with twice replacement of buffer. This dialysate was used for assay.

Enzyme assay:

- ❖ 1 ml of 0.05 M catechol and 4.5 ml of 0.2 M phosphate buffer (pH 6.8) was incubated and the reaction was initiated by adding varying amounts of enzyme extract, at 30°C by making the final volume to 5 ml.
- ❖ The rate of increase in absorbance was measured for every 30 second up to 3 minutes against the blank (which is prepared in the absence of enzyme) at 410 nm.
- ❖ The change in absorbance between 30 to 180 second of incubation was plotted in graphical form and the enzyme activity was calculated from the linear part of the curve.
- ❖ Express the enzymatic activity as unit/min at 410 nm considering one enzyme unit as the change in absorbance of 0.001/min and specific activity as unit/min/mg protein.

Estimation of Catalase (CAT) enzyme: Catalase activity was assayed following the method of **Luck (1974)**

Principle:

The UV absorption of hydrogen peroxide can be measured at 240 nm, whose absorbance decreases when degraded by the enzyme catalase. From the decrease in absorbance, the enzyme activity can be calculated.

Reagents:

a) Preparation of enzyme extraction buffer (50 mM)

- ❖ Potassium hydrogen phosphate (0.1M, dissolved 1.74 g in 100 ml water)
- ❖ Potassium dihydrogen phosphate (0.1M, dissolved 1.36 g in 100 ml water)
- ❖ Mix both PHP & PDP in ratio of 84:16 (84 ml PHP + 16 ml PDP)
- ❖ Adjust the pH 7.0
- ❖ Add 7.42 mg EDTA + 2 g PVP in 100 ml potassium phosphate buffer (0.1M) and add 97 ml water and make final volume 200 ml.
- ❖ Take 500 mg leaf sample and homogenized in ice cold buffer with mortar and pestle.
- ❖ Filtrate the homogenized sample through four layer cheese cloth.
- ❖ Centrifuged at 4°C for 20 min at 15000 rpm.
- ❖ Collect the supernatant.

b) Preparation of reagents for catalase activity

- ❖ Potassium hydrogen phosphate (0.1M, dissolved 1.74 g in 100 ml water)
- ❖ Potassium dihydrogen phosphate (0.1M, dissolved 1.36 g in 100 ml water)
- ❖ Mix both PHP & PDP in ratio of 61:39 (61 ml PHP + 39 ml PDP)
- ❖ Adjust the pH 7.0

- H_2O_2 (6.66mM) 13.6 μl of 30% H_2O_2 was diluted in 20 ml distilled water.

Procedure:

- Take 1 ml H_2O_2
- Take 2 ml already made buffer
- Take 100 μl enzyme extract (diluted 5 times with water, 1 volume of extract + 4 volume of water)
- Take the observations under spectrophotometer at 240 nm.
- Take the reading at 0, 15, 30, 45 seconds and 1 minute.
- Express enzyme activity as unit/min and specific activity as unit/min/mg protein present in the sample.

Estimation of Phenyl-alanine Ammonia Lyase (PAL): Phenylalanine ammonium lyase catalyses deamination of phenylalanine to trans-cinnamic acid. In diseases resistance mechanism, the enzyme plays an important role in the conversion of phenylalanine to coumaric acids.

Principle:

The enzyme may be assayed by measuring the appearance of trans-cinnamic acid from phenylalanine spectrophotometrically at 290 nm.

Reagents :

- Sodium borate buffer
- 2-mercaptoethanol
- Polyvinylpyrrolidone

Procedure: The l- phenylalanine ammonia lyase activity was determined as the rate of conversion of l- phenylalanine to trans-cinnamic acid as described by **Dickerson et al. (1984)**.

- One gm of leaf sample was homogenate in 5 ml of ice cold 0.1 m sodium borate buffer, pH 7.0, containing 1.4 mM 2-mercaptoethanol and 0.1 g insoluble polyvinylpyrrolidone.
- The extract was filtered through cheese cloth and filtrate was centrifuged at 20000g at 4°C for 30 minutes.
- The supernatant was used as enzyme source.
- An aliquate of 0.8 ml enzyme extract was incubated with 1 ml of 0.12 M l- phenylalanine in 0.1 M borate buffer for 30 minutes at 30°C.
- The absorbance value was recorded at 290 nm.
- The amount of trans-cinnamicacid formed was calculating using standard curve of cinnamic acid.
- Enzyme activity was expressed as mol trans-cinnamic acid produced min^{-1} mg^{-1} protein.

Estimation of Peroxidase (POD): Peroxidase occurs in plants and animals cells. It is ubiquitous in nature. It catalyses the oxidation of many organic compounds by hydrogen peroxide. Increase in peroxidase activity has been reported in a number of host-parasitic interactions.

Principle:

The enzyme activity is assayed using o-dianisidine as hydrogen donor and H_2O_2 as electron acceptor. The rate of formation of yellow range coloured dianisidine dehydrogenation product is a measure of the POD activity and can be assayed spectrophotometrically at 430 nm.

Reagents:

- 0.1 M Potassium phosphate buffer, pH 6.0.
- 0.01 M o-Dianisudine in methanol: Prepared fresh.
- 0.02 M H_2O_2 (20 mM): Dilute 0.227 ml of 30% H_2O_2 to 100 ml with water. Prepared fresh.
- 2N H_2SO_4. (Add 5.5 ml in 100 ml water)

Preparation of enzyme extraction buffer (50 mM):

- Potassium hydrogen phosphate (0.1M, dissolved 1.74 g in 100 ml water)
- Potassium dihydrogen phosphate (0.1M, dissolved 1.36 g in 100 ml water)
- Mix both PHP & PDP in ratio of 84:16 (84 ml PHP + 16 ml PDP)
- Adjust the pH 7.0
- Add 7.42 mg EDTA + 2 gm PVP in 100 ml potassium phosphate buffer (0.1 M) and add 97 ml water and make final volume 200 ml.
- Take 500 mg leaf sample and homogenized in ice cold buffer with mortar and pestle.
- Filtrate the homogenized sample through four layer cheese cloth.
- Centrifuged at 4°C for 20 min at 15000 rpm.
- Collect the supernatant.

Enzyme Assay:

- 1 ml of o-dianisudine was pipetted out into a test tube to which 0.5 ml of H_2O_2, 1 ml of phosphate buffer and 2.4 ml of distilled water were added.
- The whole mixture was incubated at 30°C and the reaction was initiated by adding 0.2 ml (400 μl water + 200 μl enzyme)of enzyme.
- After 5 minutes, the reaction was stopped by adding 1 ml of 2N H_2SO_4.
- For the blank was H_2O_2 excluded retaining all the reagents as such but additional volume of water was added.
- Take the reading at 0, 15, 30, 45 seconds and 1 minute.
- The absorbance was measured at 430 nm.

Estimation of Superoxide Dismutase (SOD):

SOD was assayed according to the method of **Kakkar et al. (1984).**

Principle:

The assay of SOD is based on the inhibition of the formation of NADH-phenazinemethosulphate-nitroblue tetrazolium formazon. The colour formed at the end of the reaction can be extracted into butanol and measured at 560 nm.

Reagents:

- Sodium pyrophosphate buffer (0.025 M, pH 8.3)
- Phenazinemethosulphate (PMS) (186 μM)
- Nitroblue tetrazolium (NBT) (300 μM)
- NADH (780 μM)
- Glacial acetic acid
- N-butanol
- Potassium phosphate buffer (50 mM, pH 6.4)

Procedure:

- The leaves samples (0.5 g), were ground with 3.0 ml of potassium phosphate buffer, centrifuged at 2000 g for 10 minutes and the supernatants were used for the assay.
- The assay mixture contained 1.2 ml of sodium pyrophosphate buffer, 0.1 ml of PMS, 0.3 ml of NBT, 0.2 ml of the enzyme preparation and water in a total volume of 2.8 ml.
- The reaction was initiated by the addition of 0.2 ml of NADH. The mixture was incubated at 30 °C for 90 seconds and arrested by the addition of 1.0 ml of glacial acetic acid.
- The reaction mixture was then shaken with 4.0 ml of n-butanol, allowed to stand for 10 minutes and centrifuged 2000 g for 10 min.
- The intensity of the chromogen in the butanol layer was measured at 560 nm in a spectrophotometer.
- One unit of enzyme activity is defined as the amount of enzyme that gave 50% inhibition of NBT reduction in one minute.

21 GENERAL INFORMATION

CONVERSION TABLES AND FORMULA

- **Glass wares**

Petri dish	Watch glass	Test tubes	Other materials
Conical/Erlenmeyer flask	Glass slides/ Cover slip	Pipettes	Inoculating needle
Beaker	Measuring cylinder	Specimen jar	Spirit lamp
	Funnel	Test tubes	Blotting paper, Non-absorbent cotton, Dropping bottle Forceps, Scissors, Glass pen marker
			Ethyl alcohol/Stains

- **Weight conversion**

1 μg	$=10^{-6}$ g
1 ng	$= 10^{-9}$ g
1 pg	$= 10^{-12}$ g
1 fg	$= 10^{-15}$ g

- **Spectrophotometric conversion**

1 A_{260}	Unit double–stranded DNA =50 μg/ ml
1 A_{260}	Unit of single–stranded DNA=33 μg/ ml
1 A_{260}	Unit of single–stranded DNA =40 μg/ml

- **DNA molar conversion**

1 μg of 1,000 bp DNA	1.52 Pmole [3.03 pmoles of ends]
1pmole of 1000 bp DNA	0.66 μg

- **Protein molar conversion**

100 pmoles of 100,000 dalton protein	=10 μg
100 pmoles of 50,000 dalton protein	=5 μg
100 pmole of 10,000 dalton protein	=1 μg

- **Protein/DNA conversion**

1 kb of DAN	=330 amino acids of coding capacity –3.7× 10^4 DaltonProtein
10,000 dalton protein	=270 bp DNA
50,000 dalton protein	= 1.35 kb DNA
100,000 dalton protein	=2.7 kb DNA

- **Abbreviations for units of measure**

°C	degree Celsius (°C = (°F - 32) X 5/9)
cm	centimetre (1 cm = 0.01 m; 2.54 cm = 1 inch)
g	gram (1 g = 0.001 kg = 0.035 ounce; 454 g = 1 lb)
ha	hectare (1 ha = 10 000 m^2 = 2.47 acres)
h	hour
kg	kilogram (1 kg = 1000 g = 2.2 lb)
km	kilometre (1 km = 1000 m = 0.62 mile)
kPa	kilopascal (1 kPa = 0.145 psi; 1 psi = 6.89 kPa)
L	litre (1 L = 1000 mL = 0.88 quarts)
m	metre (1 m = 100 cm = 39.4 inches)
min	minute
mL	millilitre (1 mL = 0.001 L)
mm	millimetre (1 mm = 0.001 m)
μm	micrometre (1 μm = 10^{-6} m)
mS	millisiemens (unit of electrical conductivity; 1 mS = 1 mmho)
ηm	nanometre (1 ηm = 10^{-9} m)
ppm	parts per million
t	tonne (1 t = 1000 kg = 2205 lb)

- **Conversions from per cent to ppm and dose per ha to per kg soil and per m2**

Percent to ppm

1 per cent	=10,000 ppm	= 10 g per litre or kg
0.1 per cent	=1000 ppm	= 1000 mg per litre
0.01 per cent	= 100 ppm	= 100 mg per litre
0.001 per cent	= 10 ppm	= 10 mg per litre
0.0001 per cent	= 1 ppm	=1 mg per litre

Dose per ha to per acre, per kg soil and per m2

X kg per ha	=	2X/5 kg per acre
X kg per ha	=	X/2 mg per kg soil
X kg per ha	=	X/10 g per m^2

PLANT PATHOLOGY RELATED WEB SITES:

- www.apsnet.org
- www.bspp.org.uk
- www.ipsdis.org
- www.ifgb.uni_hannover.de
- www.cabr.org/institute/imi/grc/ htm
- www.bspp.org.uk/mppol
- www.scisoc.org
- www.phytoannulreview.org
- www.biology.amu.edu.au/groups/mes/video
- www.vms.utexas.edu/~jdana/history/famine.html
- www.helios.bto.ed.ac.uk/bto/microbes.htm
- www.tulane.edu/~dmsander/garryfavweb.html
- www.image.fs.uidaho.edu/vide/
- www.micro.msb.le.ac.uk/335/viorids/html
- www.bidogi.e.,uni-hamburg.de/bonline/e35/35.htm
- www.hcs.ohio-state.edu/hcs300/bact.htm
- www.wheat.pw.usda.gov/~lazo/docs/xmal
- www.nematode.unl.edu/
- www.wcrl.ars.usda.gov/cec/teaching/nema.htm
- www.botit.botany.wisc.edu/toms_fungi/
- www.masfungi.org/
- www.britmycolosoc.org.uk/
- mycology.cornell.edu/
- www.mycolog.com/fifthtoc.html
- www.suddenoakdeath.org/
- www.helios.bto.ed.ac.uk/bto/microbes/biotroph.htm
- www.zoosporic-fungi.dmc.maine.edu/
- www.science.siu.edu/parasiticplants/index.html
- www.rms.nau.edu/mistletoe
- www.cropsci.uiuc.edu/faculty/gca/kaffe/index.htm
- www.scri.sari.ac.uk/MBN/Erwinia/chartxt.htm
- www.cnr.umn.edu/FR/extension/foresthealth/foreststandshadetreehealthpage.htm
- www.aces.edu/dept/plantdiagnosticlab
- www.cas.psu.edu/docs/CASDEPT/PLANT/ext/bact_dis.html
- www.ca.uky.edu/agcollege/plantpathology/PPAExtesnion/Ppalinks.htm
- www.ohioline.osu.edu/
- www.sciencemedcentral.com/bpo/general/home.htm
- www.agdia.com/

IMPORTANT SYMPTOMS OF PLANT DISEASES

Mildew	is seen as white, gray, brownish or purplish growth on the host surface. In case of downy mildew, the superficial growth is a tangled cotton, or downy growth while in powdery mildews, enormous number of spores are formed on superficial growth of the fungus giving the host surface a dusty or powdery appearance. Black minute fruiting bodies may also develop in the powdery mass e.g. downy mildews of maize and pearl millet, powdery mildews of pea and cucurbits	 
Rots	Due to these diseases infection the parenchymatous, collenchymatous tissues and pith are killed. On the basis of plant organ affected, the rots are designated as root rots, foot rots, collar rots, neck rots, bud rots, fruits rots, etc. On the basis of colour, such as white, red, black and brown rots have also been designated. The firmness of affected tissues determines whether it is soft rot or dry rot for e.g. soft rot of fleshy vegetables, dry rot of potato, etc.	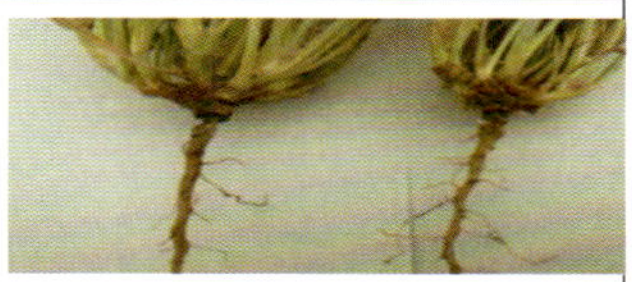 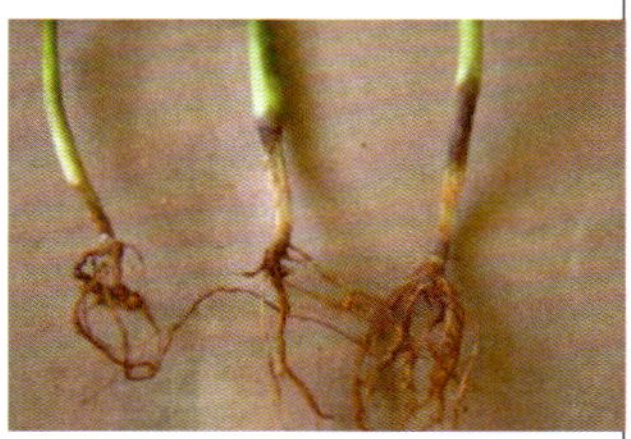
Rusts	The important symptoms of this disease as rusty. The rusts appear as relatively small pustules of spores, usually breaking through the host epidermis. The pustules may be either dusty or compact and red, brown or black in colour e.g. black, orange and yellow rust of wheat, pearl millet rust, linseed rust, etc.	

Smuts	The smut means a sooty or charcoal-like powder. In plant diseases known as smut affected parts of the plant show a black or purplish-black dusty mass. These symptoms usually appear on floral organs, particularly the ovulary, and the pustules are usually considerably larger than those of the rusts. These symptoms may also be found on stem, leaves as well as roots e.g. smuts of wheat and pearl millet, sorghum smuts and covered smut of barley.	
Canker	The localized corky necrosis on any part of the shoot or the fruit. It is generally formed in perennials when invaded by fungal or bacterial pathogen e.g. citrus canker	
Scab	Localized lesions devolve on fruits, leaves, tubers, etc. The lesions may be slightly raised or sunken and cracked. The necrosis is usually superficial and restricted mainly to a few layers of the cells e.g. Apple scab.	
Blight	This is rapid browning and death of leaves, branches, twigs and floral parts when invaded of pathogens e.g. bacterial blights of rice, cluster bean and moth bean, leaf blight of maize, Alternaria blight of tomato, Alternaria blight of cumin, etc.	
Blast	Leaf blade, bud, other plant parts or the whole plant is rapidly killed e.g. blast of rice.	

Anthracn ose	Ulcer like necrotic and sunken lesions on stem, leaf, fruit or flower e.g. Anthracnose of cotton and chillies etc.	
Streaks and stripes	Small linear lesion develops on leaf blade, leaf sheath, stem and other parts. The streak may retain limited or may enlarge on spread to make stripes. The colour of streaks and stripe may be yellow, brown, black, etc. e.g. stripe disease of barley, wheat streak, bacterial streak of rice and maize streak.	
Spots	Localized lesion consisting of dead and collapsed cells on the host. They may develop on stem, leaf and fruit. Eg. leaf spot of rice and brown spot of maize.	
Die back	As the name indicates such diseases are characterized by drying of plant organs, especially stem or branches from the tip backwards. eg. die back of citrus and chillies.	
Damping off	Tissues in the basal portion of stem or crown region are affected leading to the death of seedlings in the nursery or field e.g. Damping off of vegetable seedlings.	

Wilts	These diseases are characterized by drying of the entire plant. The leaves and other green or succulent parts lose their turgidity, become flaccid and droop. Black or brown discolorations in vascular tissues are characteristically observed in wilted plants. The vascular system of root/stem is affected and the plant dries up as though the water supply has been cut off e.g. Wilts of cotton, cumin, tomato, pigeon pea, linseed, etc.	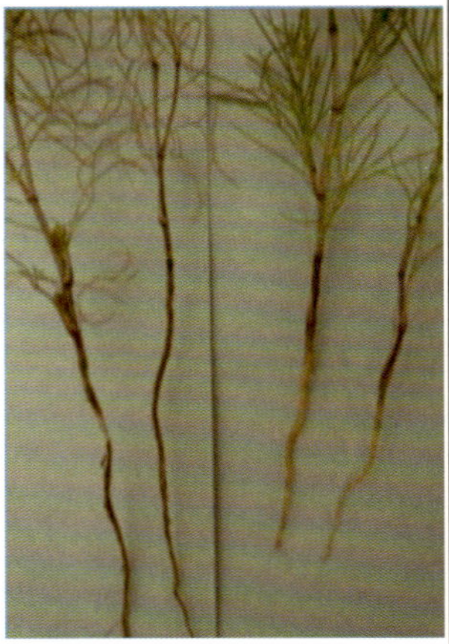
White blisters	On leaves of cruciferous and other plants there may be found numerous white blister-like pustules which break upon the epidermis and exposé powdery masses of spores e.g. white rust of mustard.	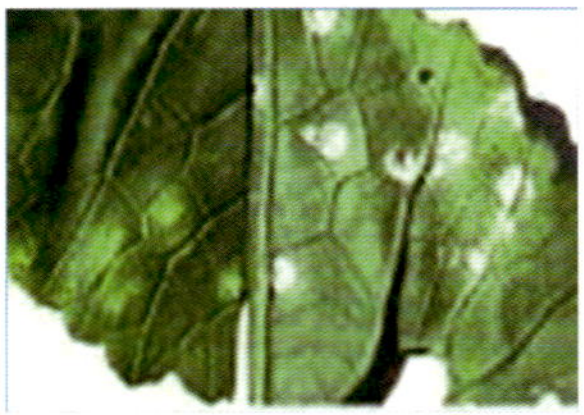

Gall	A tumor formed by hyperplasic growth of unorganized cells or localized enlargement of plant. e.g. club root of cabbage, wart of potato, stem gall of coriander	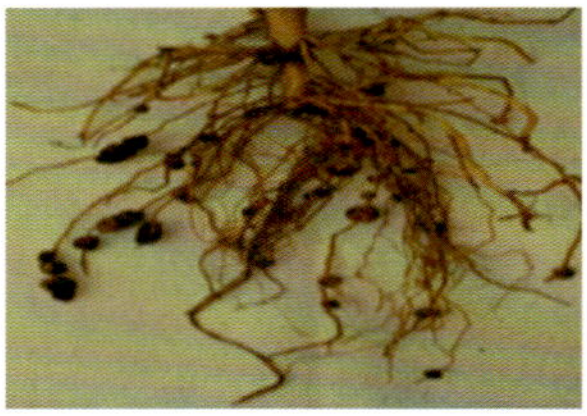
Gummosis	Water-soaked patches may develop on stem near the ground which turns brown. The bark splits and gum like ooze exudes e.g. Citrus gummosis.	

Term	Description	Illustration
Mosaic	One of the most common and obvious effects of virus infection is the development of patterns of light and dark green or yellow and green mottling of the foliage. This pattern varies with the virus-host interaction e.g. Yellow mosaic of moong and moth bean, bhindi yellow vein mosaic, potato mosaic, cowpea mosaic, cucumber mosaic, etc.	
Mottle	It is a closely related term of mosaic, and an irregular pattern of indistinct light and dark areas.	
Yellows	When the entire leaves are uniformly yellows because of almost uniform reductions of chlorophyll even entire leaf surface e.g. beet yellows and yellows of Vinca rosea, etc.	
Chlorosis	It is the general term used to denote destructions or inhibition of formation of chlorophyll subsequent to infection and occurs only when young growing leaves are infected e.g. aster yellows	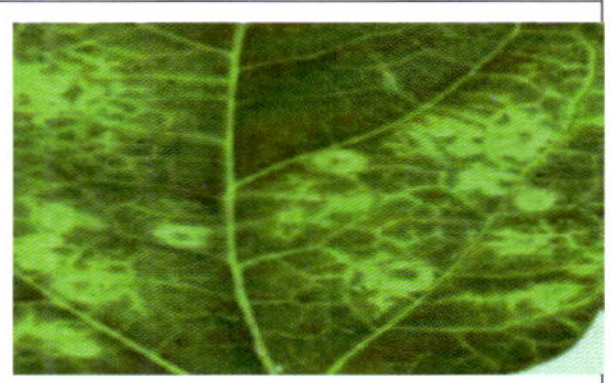
Vein-clearing	Normal green colour of veins of leaf is lost due to lack of chlorophyll but interveinal areas remain green e.g. bean mosaic.	
Vein-banding	When chlorotic tissue occurs along the veins in a regular manner in the form of chlorotic bands	

Vein-mosaic	Implies the presence of chlorosis along the main veins and unlike vein-chlorosis, the lighten areas are irregular in distribution and shape and may also cover the adjacent interveinal tissue in the form of irregular patterns	
Chlorotic lesions	These are formed when the cells in localized areas round in the site of infection, loose their chlorophyll and other pigments. The lesions may be pale to almost white e.g. Sugar beet mosaic virus	
Necrotic lesions	The affected leaf area become characterized by dead cells, the extent of necrosis varying in size and shape from a small pinhead to as large on 5 mm in diameter. The lesions may appear chlorotic in the beginning but they turn necrotic e.g. tomato spotted wilt virus in tobacco.	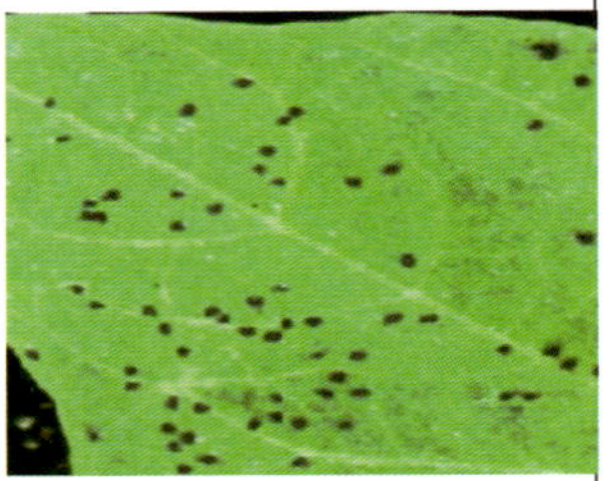
Ring spot	The type of lesion are formed when concentric group of chlorotic or necrotic and normal green tissues alternate are those formed due to tobacco ring spot virus.	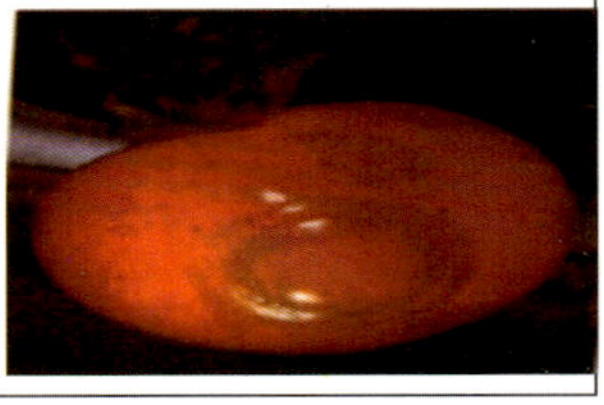
Hypertrophy	It is characterized by the over growth due to excessive cell enlargement and is noticed in diseases like potato wart, crown gall, root knots, etc.	
Hyperplasia	It means under development of tissues. The stunting and dwarfing in plants is seen in diseases like dwarf bunt of wheat, little leaf of brinjal, etc.	

Malformation	Viruses commonly alter the morphogenesis of infected plants and cause many abnormalities in their growth and form.	
Tumors	Tumors are large and irregular out-growth of tissue formed by abnormal increase in number and size of cells can grow autonomously and may sometimes show a slight differentiation of tissue e.g. wound tumor virus.	
Leaf curl	Thickening, curling and distortions of leaves e.g. chillies leaf curl, peach leaf curl, papaya leaf curl, tomato leaf curl, cotton leaf curl, etc.	
Witches-broom	It is a special type of organoid malformations in which a large number of slender branches arise from a localized area and run upward more or less parallel to each other giving the appearance of a broom or the leaves become reduced in size, the internodes are shortened and the abnormal growth of leaves results in a density-packed structure called witches broom e.g. potato witches-broom, legume witches-broom, etc.	 
Phyllody	It is a disease syndrome in which floral parts are converted into leafy or almost leaf-like structure and mostly occurs in sepals and petals, and rarely in stamens and carpels e.g. sesamum phyllody.	

Leaf rolling	In this case, upward or downward rolling of leaves is observed e.g. Potato leaf roll.	
Yellow mosaic	Intermingling between normal green and yellow colour of leaf in irregular manner e.g. yellow mosaic of moong bean and moth bean.	

TERMINOLOGY

- **Abiotic disease-** disease or disorders caused by a non-living or non replicableentity, such as water or nutrient stress, wind, frost or hail.
- **Acervulus** (pl. Acervuli) - a cushion-like structure bearing asexual conidia of fungi in the Deuteromycete Order Melanconiales. Larger and more organized than a sporodochium, formed most commonly embedded in host tissue.
- **Acropetal -** spore produced at the top of a chain of spores.
- **Aeciospore -** dikaryotic spore produced by rust fungi that infects the alternatehost species in heteroecious rust species; produced in a structure called an aecium (pl. aecia)
- **Aflatoxin -** a highly toxic mycotoxin produced by Aspergillus fumigatus on peanuts and other crops.
- **Alternate host -** a second host required for the completion of the life cycle of aplant pathogen (e.g. Rust fungi).
- **Alternative host -** a secondary host (often one of many) that may be colonizedby a plant pathogen to complete part of the disease cycle (e.g. weeds asoverwintering hosts for viruses). Not required for completion of the life cycle
- **Anamorph -** asexual stage of Ascomycete fungi associated with production of conidia; classified in the Deuteromycetes.
- **Anastomosis -** The fusion of two cells from different individuals to a form a single cytoplasmic union.
- **Antheridium -** male gametangium of many fungi.
- **Apothecium** (pl. Apothecia) - often large, fleshy, stalked disc or cup shaped fruiting body of the ascomycetes bearing asci and ascospores which are actively discharged and dispersed aerially.
- **Appressorium** (pl. Appressoria) - swelling at the end of a germ tube or hypha of plant pathogenic fungi that adheres to the cuticle and gives rise to thepenetration peg which penetrates the cuticle and epidermal cell wall.
- **Aseptate -** Hyphae with no cross walls; coenocytic; typical of lower fungi
- **Ascomycete -**fungi that produces septate haploid mycelium, and sexual sporesin an eight-spored sac-like ascus (ascospores) and asexual conidia.
- **Ascospore -** sexual spores of the ascomycetes produced after one meiotic and one mitotic cell division of a diploid zygote nucleus. Usually produced in groups of 8 inside an ascus.
- **Ascus** (pl. asci) - sac-like structure containing (usually) eight ascospores. Characteristic of the ascomycetes
- **Autoecious -** rust fungi that complete their life cycle on one host.
- **Basidiomycete -** fungi with septa, often dikaryotic, mycelium that produce sexual haploid basidiospores.

- **Basidiospore -** sexual haploid spores of basidiomycetes produced on a club-shaped basidium after one meiotic division of a diploid zygote nucleus.
- **Basidium** (pl. basidia) - club-shaped cell that contains the diploid zygote nucleus of the basidiomycetes and which gives rise to four haploid basidiospores after a single meiotic division.
- **Basipetal -** spore produced at the bottom of a chain of spores
- **Biotic disease -** A disease casued by a living or otherwise replicable entity such as fungi, bacteria, mycoplasmas and viruses.
- **Biotroph** - A parasite that requires living host tissue as a source of nutrients.
- **Blight -** Complete necrosis of above-ground plant parts.
- **Bordeaux mixture -** The name of the first fungicide discovered, by chance, to control downy mildew of grapes in 1885.
- **Chlamydospore -** Asexual, thick-walled, single-celled, overseasoning spore. Chlorosis - yellowing of host tissue.
- **Chytridiomycete -** simple fungi consisting of little more than a sporangium and sexual spores and restricted aseptate mycelium. Produce motile zoospores, cause several root rot diseases and act as virus vectors.
- **Clamp Connections -** "clamp-like" fingers joining adjacent cells in Basidomycete hyphae to maintain the dikaryotic nature of the cells in the mycelium
- **Class -** The taxonomic level below Phylum or Division and above Family.
- **Cleistothecium** (pl. Cleistothecia) - closed, sexual fruiting body of the Ascomycetes containing asci and ascospores adapted as overwintering structure. (e.g. Powdery mildews).
- **Coenocytic -** Hyphae with no cross walls; aseptate; typical of lower fungi (see Aseptate, Septum)
- **Conidium** (pl. conidia) - asexual spores of the Deuteromycetes and Ascomycetes produced after mitotic cell divisions of specialized cells formed ON the mycelium without sexual processes.
- **Cyst -** non-motile phase of zoospore prior to germination.
- **Damping off -** Disease resulting in the death of seeds, seedlings and young plants.
- **Deuteromycete** - fungi with septate mycelium bearing only asexual spores produced by mitosis (conidia), often asexual stages of Ascomycetes.
- **Dikaryon** - Somatic cells of a fungus that contain two nuclei.
- **Disease** - any disturbance of a plant that interferes with normal growth and development, economic value or aesthetic value; a continuous, intermittent or progressive condition as opposed to an injury.
- **Disease Cycle** - sequential stages of disease due to the different life stages of the pathogen and responses of the host to disease.
- **Dissemination** - Spread of pathogen inoculum. Part of the disease cycle.
- **Division** - The taxonomic level below Kingdom and above Class. Reserved for the classification of plants.
- **Epinasty** - downward bending of leaves and petioles without loss of turgor.

- **Facultative Parasite** - Organism able to survive indefinitely as saprophyte but capable of causing plant disease (e.g. *Fusarium*)
- **Facultative Saprophyte** - Organisms which cause plant disease but with the ability to survive indefinitely as saprophytes (e.g. Venturia inaequalis).
- **Gametangium** - cell containing haploid gamete nuclei
- **Germ Tube** - the specialized hypha that issues from a fungal spore or other propagule upon germination.
- **Haustorium** (pl. Haustoria) - specialized feeding organ formed by many biotrophic fungi that invaginates the host plasma membrane and forms an intimate interface between the host and pathogen through which nutrients pass.
- **Heteroecious** - rust fungi that require two alternate hosts to complete their life cycle.
- **Heterothallic** - two mating types required for sexual reproduction. Self-sterile or self incompatible individuals.
- **Holomorph** - The totality of all stages of the life cycle of an Ascomycete fungus including all sexual and asexual spore stages; classified in the Ascomycetes and takes the same name as the teleomorph.
- **Homothallic** - self-fertile individuals; sexual reproduction occurs on the samethallus.
- **Hyperplasia** - abnormal increase in cell division.
- **Hypbertrophy** - abnormal cell enlargement.
- **Hypha** (pl. hyphae) - filamentous, vegetative body of a fungus.
- **Incubation period** - The time between inoculation and the appearance of disease symptoms.
- **Indirect germination** - germination of a sporangium by zoospores.
- **Karyogamy** - fusion of two nuclei.
- **Latent Infection** - Asymptomatic infection of a host until symptom expression is triggered by a specific physiological event such as ripening (*Colletotrichum musae* on banana).
- **Lesion** - A discrete symptom of disease. Often a spot, blotch or other identifiable site that results from infection.
- **Life cycle** - the morphological stage or stages of an organism, each serving one or many ecological functions, which appear sequentially as the organisms grows, develops and reproduces.
- **Macroconidium** (pl. Macrocondia) - large multicelled conidia produced by Deuteromycete fungi that produce more than one type of conidia.
- **Macrocyclic** - rust fungus disease cycle which includes two alternate hosts.
- **Meiosis** - The "reduction" division, whereby a diploid or polyploid cell undergoes a sequence of two divisions that results in the daughter cell having (usually) half the number of chromosomes of the mother cell. Often associated with sexual reproduction
- **Microconidium** (pl. Microconidia) - small, usually single-celled, conidia produced by Deuteromycete fungi that produce more than one type of conidia.rust fungus disease cycle this includes only one host.

- **Mitosis -** A vegetative cell division that results in daughter cells that are clones (i.e. identical) to the mother cell. Both mother and daughter cells have the same number of chromosomes.
- **Monocyclic -** Disease epidemic caused by pathogens that complete only one disease cycle in one year (e.g. smuts, soil borne pathogens, vascular wilt pathogens).
- **Mycelium** (pl. Mycelia) - filamentous, vegetative body of a fungus consisting of many hyphae.
- **Mycoplasma -** Wall-less prokaryotes (bacteria) that live in the phloem of the host.
- **Mycotoxin -** a substance produced by a fungus that is toxic to humans and animals. (See Aflatoxin)
- **Myxomycete -** slime molds, ameboid cells form a plasmodium (mass of ameboid cells) which can move around and form sporangia, motile zoospores and more amoebae (cells which move by protoplasmic streaming). Not true fungi.
- **Necrosis -** death of host cells due to biotic or abiotic factors pathogen resulting in collapse and discoloration of tissue.
- **Necrotroph -** A parasite that lives on dead host tissue after killing the tissue with enzymes and toxins.
- **Obligate parasite -** A parasite that requires living host tissue as a source of nutrients.
- **Oogamy -** Oogamous fertilization in which a large female gametangium (oogonium or egg cell) is fertilized by a small male gametangium (antheridium)
- **Oogonium -** female gametangium of the oomycetes.
- **Oomycete -** alga-like fungi that have diploid, aseptate hyphae, reproduce sexually by oospores and have asexual sporangia often containing motile zoospores.
- **Oospore -** sexual, overwintering, spore of Oomycete fungi. Product of oogamous fertilization.
- **Ostiole -** opening or pore in wall of closed fruiting bodies of Ascomycetes (Perithecium) and Deuteromycetes (Pycnidium) which allows active or passive dispersal of conidia from the fruiting body under suitable conditions. Breakdown of the fruiting body is not required for dispersal of spores.
- **Parasite -** An organism living in or on another living organism and obtaining nutrients from it, without necessarily giving benefit to, or causing damage to the host.
- **Pathogen -** an organism or other replicable entity that damages a host plant thereby causing disease.
- **Pathogenicity** - qualitative ability of a plant pathogen to cause disease
- **Perithecium** (pl. Perithecia) - closed, ostiolate fruiting body of the Ascomycetes bearing asci and ascospores. Ascospores are active and passively dispersed by water and wind.

- **Phylogeny** - The evolutionary or ancestral history of organisms.
- **Phylum** - The taxonomic level below Kingdom and above Class. Reserved for the classification of animals and fungi.
- **Phytoalexin** - literally "plant warding-off" compound. A toxic compound produced by the plant to kill or repel pathogens.
- **Plasmodiophoromycete** - simple fungi causing club-root of crucifers. Amoeboid stage, similar but unrelated to the Myxomycetes and motile zoospores are produced.
- **Plasmogamy** - fusion of two cells.)
- **Polycyclic** - Disease epidemic caused by pathogens that complete many disease cycles in one year, often as a result of secondary inoculum (e.g. rusts, powdery mildews, downy mildews, Rhizopus).
- **Primary inoculum** - Type of inoculum that survives adverse conditions, overwinter or oversummer, and causes the first infections in the next growing season.
- **Propagule** - a spore or other germinable structure, such as a sclerotium that can serve to propagate a pathogen.
- **Pseudothecium** (pl. Pseudothecia) - closed ostiolate, stromatic fruiting body of the Ascomycetes bearing asci and ascospores. Ascospores are dispersed actively and passively by wind and water.
- **Pycnidium** (pl. Pycnidia) - asexual, closed, ostiolate fruiting body of the Deuteromycetes, specialized for overseasoning and dispersal of conidia by water and wind.
- **Rhizomorph** - A "root-like" aggregation of hyphae, that is capable of conducting water and growing toward a food source in an organized fashion.
- **Saprophyte** - an organism that obtains its nutrients from dead organic matter.
- **Secondary inoculum** - typically asexual spores of a pathogen acting as inoculum responsible for in-season and in-field spread of a disease after it has been established by the primary inoculum.Inoculum responsible for polycyclic disease epidemics.
- **Septum** (pl. septa) - Specialized cross walls in hyphae of higher fungi, separating hyphae into cells.
- **Sclerotium** - Asexual, thickwalled, multicelled, overseasoning structure.
- **Sign** - The name given to the appearance of fruiting bodies and soma of the pathogen itself on the surface of a diseased plant.
- **Soil Inhabitant** - an organism that grows and reproduces in the soil environment
- **Soil Invader** - an organism that exists in the soil only as a resting overseasoning propagule and does not grow or reproduce in the soil
- **Soma** (pl. Soma) - vegetative body of fungi; mycelium; hyphae.
- **Species** - The taxonomic ranking immediately below genus that includes morphologically and genetically similar individuals that freely interbreed.
- **Spermatia** (sing. spermatium) - haploid spore produced in a spermogonium (pl. spermogonia) which fertilizes the receptive hyphae in a compatible

spermogonium and leads to the production of dikaryotic hyphae.

- **Sporangiospore -** asexual spores produced inside a specialized cell (sporangium). (see Sporangium).
- **Sporangium** (pl. Sporangia) - Sack-like structure containing asexual spores (Sporangiospores) of lower fungi.
- **Sporodochium** (pl. Sporodochia)- asexual, cushion-like fruiting body of Deuteromycete subclass Hyphomycetidae, adapted for water or splash dispersal of spores. Smaller and less organized than acervuli.
- **Sterigma** (pl. Sterigmata) - short projection or hyphal branch which supports, and subtends, a sporangium, conidium or basidiospore.
- **Stroma -** dense, tissue-like, fungal material often associated with fruiting bodies such as pseudothecia.
- **Symbiont -** an organism living on or in another organism from which both organisms derive mutual benefit.
- **Symptom -** outward expression of disease by tissues of the host.
- **Synemma** (pl. Synemmata) - asexual fruiting body, bearing conidia, that consists of an aggregation of hyphae forming a long stalk which raises the sporogenous head well above the substrata for dispersal of conidia by wind, water and insects.
- **Teleomorph -** stage of an Ascomycete fungus that represents the sexual stage; Classified in the Ascomycetes.
- **Teliospore -** the specialized probasidium (immature basidium) of rust fungi adapted specifically for overseasoning and which germinates by producing a club-shaped metabasidium (mature basidium) bearing four haploid basidiospores.
- **Thallus -** vegetative body of a fungus; mycelium; hyphae; soma.
- **Tylosis** (pl. tyloses) - An overgrowth of xylem parenchyma into the xylem vessel in response to infection by a vascular pathogen of plants.
- **Urediniospore** - Uredospore
- **Uredospore -** the dikaryotic, asexual "summer" spore of rust fungi. Acts as secondary inoculum of the rust fungi and is often a rusty-orange color. Produced in the structure called a uredium (pl. uredia). Capable of reinfecting the host on which it is produced.
- **Virulence -** The degree of pathogenicity of a given pathogen. Often a quantitative term when comparing isolates in a given population. Quantitative term.
- **Witches' Broom** - excessive proliferation of weak shoots or roots, rather like a straw-broom, arising from a bud or other meristematic region caused by fungi, mites, viruses or Mycoplasmas.
- **Wilt -** loss of turgor resulting in flaccidity of host tissue.
- **Yeast -** a fungus that completes all or most of it's life cycle as a single cell.
- **Zoospore -** asexual, motile spore of lower fungi.
- **Zygomycete -** fungi with aseptate mycelium which produce sexual zygospores and asexual sporangiospores in sporangia.
- **Zygospore -** Sexual, overwintering, spore of zygomycetes resulting from fusion of two gametangia.

SUGGESTED READINGS

- Laboratory manual of microbiology, Biochemistry and Molecular biology-Jyoti Saxena, Mamta Baunthiyal and Indu Ravi, 2012, Scientific Publishers (India)
- Detection of Pathogens and Their Management-J.P. Verma, A.Verma & D. Kumar, 1995, Angkor Publ. N.Delhi.
- Plant Pathology -G.N. Agrios,2006, Academic Press, New York
- The Essentials Of Viruses, Vectors And Plant Diseases -A.N. Basu & B.K. Giri,1992, Wiley Eastern Ltd., N. Delhi
- Biotechnology in Plant Diseases-I.Chet, 1993, John Wiley & Sons. Inc. New York
- Molecular Plant Pathology, A Practical Approach Vol. II-S.J. Gurr, M.J. Pherson & D.J. Bowles,1992, Oxford University Press, U.K.
- Methods In Plant Virology- Stephen A.Hill , 1984, Blackwell Scientific Publication, Oxford London.
- Molecular Plant Pathology- Dickinson,M., 2008, Bios
- Diagnosis of Plant Viruses and Allied Pathogens-Y.S.Ahlawat, 2010, Studium Press
- Methods In Phyto- bacteriology -Z. Klement, K.Rudolph & D.C. Sand, 1989, Akademial Kiado, Hungary
- Agrios, G.N. 2005. *Plant Pathology* Vth edn. Academic Press, New York.
- Brunt, A.A., Crabtree, K., Dallwitz, M.J., Gibbs, A.J., Watson, L. and Zurcher, E.J. (eds) (1996 onward) Plant Viruses on Line: description and lists from the VIDE database. http://biology.anu.edu.au/groups/mes/vide/
- Dasgupta, MK. 1998. Phytonematology. Naya Prakash, Naya Udyog, Bidhan Sarai, Kolkata, India.
- Gupta, VK and Sharma, Satish, K. (eds.) 2000. Diseases of Fruit Crops. Kalyani Publishers, New Delhi.
- Johnson, R. and Jellis, GJ (eds.) 1993. Breeding for Disease Resistance. Kluwer, Hingham, Massachusetts.
- Joshi, L.M., Singh, D.V., and Srivastava, K.D. 1986. Problem and progress of wheat pathology in South Asia. Malhotra Publishing House, New Delhi
- Mehrotra, RS. 2003. *Plant Pathology*. Tata McGraw Hill, New Delhi.
- Rangaswami, G. 1998. Diseases of crop plants in India,3rd Edn. Prentice-hall of India, New Delhi.
- Raychaudhuri, SP, and Verma Jeev, P. (eds) 1984-1990. Vol. I. Diseases of Cereals, Vol.II. Diseases of Fruits, Vol.III. Review of Tropical Plant Pathology (Vegetable Diseases, Vol IV. Diseases of Plantation Crops and Forest Trees Vol. V. Diseases of Fibre and Oil Seed Crops. Today & Tomorrow Printers and Publishers, New Delhi.
- Singh, R.S. 1998. Plant diseases, 7th Edn. Oxford & IBH Pub. Co. Pvt. Ltd. New Delhi
- Singh, U.S., Mukhopadhyay, A.N., Kumar, J. and Chaube, H.S. 1992. Plant diseases of International importance, Vol 1-4, Prentice Hall, Eaglewood cliffs, New Jersey.

- Smith, K.M. 1972. A text book of plant virus diseases. Academic Press, New York.
- Timmer, L.W., Garnsey, S.M. and Graham, J.H. 2000. Compendium of citrus diseases 2nd edn. APS Press, St. Paul, MN.
- Vanderplank, JE. 1984. Disease Resistance in Plants. 2nd Edn. Academic Press, Orlando, Florida.
- Pelczar, Jr. Michel J. Chan, E.C.S. and Krieg, Noel R. 1997. Microbiology. Tata McGraw -Hill Edition, 1993. India.
- Rangaswami, G. and Bagyaraj, D.J. 1993. Agricultural Microbiology. Prentice Hall of India Pvt. Limited, New Delhi.
- Vishunavat, K. and Kolte, S.J. 2005. Essentials of Phytopathological Techniques. Kalyani Publishers, New Delhi.
- Agrios GN. 2005. Plant Pathology. 5th Ed. Academic Press, New York. (Indian Ed.)
- Mehrotra, R.S. and Aggarawal, A. 2007. Plant Pathology. Tata McGraw Hill Publishing Co. Ltd., New Delhi
- Godara, S.L., Kapoor, B.B.S. and Rathore, B.S. 2010. Disease Management of Spice Crops. Madhu Publications. Bikaner, Rajasthan.
- Alexopoulos CJ, Mims CW & Blackwell M.2000. Introductory Mycology.5th Ed. John Wiley & Sons, New York.
- Rangaswamy ,G and Mahadevan,A .2001. Diseases of crop plants in India. Prentice hall of India Pvt Ltd ND.
- Singh R S .2007 Plant Diseases.(9th Ed.) Oxford and IBH Publishing Co.Pvt .Ltd .ND
- Gupta, S.K. and Thind, T.S. 2006. Disease Problems in Vegetable Production. Scientific Publishers, Jodhpur.
- Singh, R.S. 2006. Diseases of Fruit Crops. Oxford and IBH Publishing Co. Pvt. Ltd. New Delhi.
- Cook, R.J. and Baker, K.F. 1983. Nature and Practice of Biological Control of Plant Pathogens. APS, St. Paul, Minnesota.
- Singh, R.S. 2000. Plant Disease Management. Oxford and IBH, NewDelhi.
- R.P. Singh 2013 Plant Pathology. Kalyani Publishers
- V.K. Gupta 2014 Diseases of Fruit Crops. Kalyani Publishers
- Prescott 2014 Microbiology. McGraw Hill & Co.
- Chaube H.S. Crop Diseases and Their Management. PHI
- Nene Y. L. 2015 Fungicides in Plant Diseases Control. IBH
- Vander plank, J.E. 2014 Host Pathogen Interactions in Plant Diseases. A.P.
- Janardhan, K.K. 2013 Diseases of Major Medicinal Plants. Astral
- Lakshman, H.C. 2014 Bio-fertilizers and Bio-pesticides. Pointer Publishers
- Verma, J.P. Plant Bacteriology.
- Kalyan, K. Mandal (2011) Plant Bacteriology. Kalyani Publishers, New Delhi.
- Roger, Hull (2013) Plant Virology
- Kajal kumar Biswas Plant viruses, Diseases and their management. J.K.International Publisher.

- We are acknowladge to all researcher for incorporating their photoghraphs and findings in this book.

COMPOSITION OF DIFFERENT MEDIA USED FOR FUNGUS AND BACTERIAL GROWTH

1. Czapek's- Dox agar:

Sodium nitrate (N_aNO_3)	2.0 g
Potassium dihydrogen phosphate (KH_2PO_4)	1.0 g
Magnesium sulphate ($M_gSO_4.7H_2O$)	0.5 g
Potassium chloride (KCl)	0.5 g
Ferrous sulphate)$F_eSO_4.7H_2O$)	0.01 g
Sucrose	30.0 g
Agar	20.0 g
Distilled water	1000 ml

2. Richard's media:

Potasssium nitrate (KNO_3)	10.0 g
Potassium dihydrogen phosphate (KH_2PO_4)	5.0 g
Magnesium sulphate ($M_gSO_4.7H_2O$)	2.5 g
Ferricchloride (F_eCL_3)	0.02 g
Sucrose	50.0 g
Water	1000 ml

3. Corn (maize) meal agar:

Maize	30 g
Agar	20 g
Water	1000 ml

4. Malt extract agar

Malt extract	15 g
Dipotassium hydrogen phosphate	1 g
Ammonium chloride	1 g
Citric acid N/1	15ml
Agar	20 g
Water	1000 ml

5. Potato dextrose agar

Peeled Potato	200 g
Dextrose	20 g
Agar	20 g
Water	1000 ml

6. Oat meal agar (Tuite, 1969)

Oat flakes	30 g
Agar-agar	20 g
Distilled water (to make up)	1000 ml

7. Host leaf extract agar media

Patchouli leaves	200 g
Agar -agar	20 g
Distilled water (to make up)	1000 ml

8. Sabouraud's agar

Dextrose ($C_6H_{12}O_6$)	20 g
Peptone	10 g
Agar-agar	20 g
Distilled water (to make up)	1000 ml

9. Carrot agar

Carrot	200 g
Dextrose ($C_6H_{12}O_6$)	20 g
Agar-agar	20 g
Distilled water (to make up)	1000 ml

10. V8 juice agar

V8 juice (high media)	44.3 g
Distilled water (to make up)	1000 ml

11. Asthana and Hawker's medium (Tuite, 1969)

Glucose ($C_6H_{12}O_6$)	5.0 g
Potassium nitrate (KNO_3)	3.5 g
Potassium dihydrogen phosphate (KH_2PO_4)	1.75 g
Magnesium sulphate ($MgSO_4 . 7H_2O$)	0.75 g
Agar-agar	15.0 g
Distilled water (to make up)	1000 ml

12. Brown's medium

Glucose ($C_6H_{12}O_6$)	2.0 g
L-asparagine	2 g
Dipotassium hydrogen phosphate (K_2HPO_4)	1.25 g
Magnesium sulphate ($MgSO_4 . 7H_2O$)	0.75 g
Trace element mixture	1 ml
Agar-agar	15.0 g
Distilled water (to make up)	1000 ml

13. Tochinai's agar

Potassium dihydrogen phosphate (KH_2PO_4)	0.5 g
Potassium nitrate (KNO_3)	2.0 g
Magnesium sulphate ($MgSO_4 . 7H_2O$)	1.0 g
Ferric chloride ($FeCl_3 . 6H_2O$)	Trace
Sucrose ($C_{12}H_{22}O_{11}$)	30 g
Agar-agar	15 g
Distilled water (to make up)	1000 ml

14. Martins' Rose Bengal Agar medium

Dextrose	10.0
Peptone	5.0
KH_2PO_4	1.0
$MgSO_4$	0.5
Rose Bengal	0.05
Streptomycin	0.03
Agar-agar	20.0
Distilled water	1000 ml

15. Composition of Fusarium selective medium

Peptone	15.0 g
Dipotassium hydrogen phosphate	1.0 g
Magnesium sulphate	0.5 g
Agar	15.0 g
Penta chloro nitro benzene	0.5 g
Oxgall	0.5 g
Streptomycin sulphate	0.1 g
Chlortetracycline hydrochloride	0.05 g
Distilled water	1000 ml

16. Composition of Trichoderma selective medium

Ammonium nitrate	1.0 g
Dipotassium Hydrogen phosphate	0.9 g
Magnesium sulphate	0.2 g
Potassium chloride	0.15 g
Glucose	3.0 g
Agar	15.0 g

Rose Bengal	0.15 g
Chloromphenical	0.25 g
Metalaxyl	0.30 g
Pentachloronitrobenzene	0.20 g
Distilled water	1000 ml

17. Nutrient Broth

Meat or beef extract	3 g
Peptone	5 g
NaCl	5 g
Agar	15.0 g
Distilled water	1000 ml

18. Malt Extract Yeast extract agar medium (MYE)

Yeast extract	4 g
Malt extract	10 g
Glucose	4 g
Agar	20 g
Distilled water	1000 ml

19. Mannitol salt agar

Mannitol	10.0 g
Peptone	10.0 g
Sodium chloride	75.0 g
Beef extract	1.0 g
Phenol red	0.025 g
Agar	15.0 g
Distilled water	1000 ml